GLOSSARY OF PLANT PHYSIOLOGY

GLOSSARY OF PLANT PHYSIOLOGY

By

A.S. GONTIA

SATHRUPA RAO

SSPH

SATISH SERIAL PUBLISHING HOUSE

403, Express Tower, Commercial Complex, Azadpur, Delhi-110033 (India)

Phone : 011-27672852, Fax : 91-11-27672046

E-mail :info@satishserial.com, hkjain1975@yahoo.com

Website : www.satishserial.com

Published by :

SATISH SERIAL PUBLISHING HOUSE
403, Express Tower, Commercial Complex, Azadpur, Delhi-110033 (INDIA)
Phone : 011-27672852 Fax : 91-11-27672046
E-mail : info@satishserial.com, hkjain1975@yahoo.com

ISBN 978-93-81226-24-7

Typeset at: Harminder Singh Kharb *for* Laxmi Art Creation
Printed at: Salasar Imaging Systems, Delhi-110035

PREFACE

This book is designed to frame the scattered information related to Plant Physiological terms. The glossary has been compiled keeping in view today's need for the undergraduate and postgraduate students seeking success in the various competitive examinations rendering them good jobs and would help as a ready reckoner for revision. The physiological terms have been arranged in alphabetical order and covers all the aspects of plant physiology. An attempt has been made to keep the language simple.

With great respect and reverence we acknowledge the encouragement given by our Hon'ble Vice - Chancellor Dr. G. Kalloo, Dean Faculty of Agriculture Dr. S.S. Tomar and all the faculty members of the Department of Plant Physiology. Last but not the least we would like to thank our family members for their co-operation during the preparation of the book.

Authors

CONTENTS

□□□

A

ABA - response element (ABRE)

- A six-nucleotide sequence element found in the promoters of genes regulated by abscisic acid (ABA), a plant hormone.

- A six nucleotide sequence element contained in the promoters of ABA - regulated genes, which probably binds transcriptional gene activation.

Abaxial : Located on the side furthest from the axis, e.g., the lower side of a leaf.

ABC model : Proposal for the way in which floral homeotic genes control organ formation in flowers. According to the model, organ identity in each whorl is determined by a unique combination of the three organ identity gene activities.

Abiotic :

- Not directly caused or induced by organisms Absorbance fraction of radiation incident on a surface that is absorbed.

- Referring to what is non-living. In the context of plant stress refers to stresses caused by environmental factors.

Abiotic factors : Term to denote all natural non-living forms. e.g. Air.

Abnormal seedlings : Any seedling which is not classified as a normal seedling is considered abnormal.

ABP1 gene : Gene that encodes the ABP1 protein.

ABP1 protein : A glyscosylated auxin-binding protein found in the endoplasmic reticulum (ER).

Abscisic acid, ABA :

- Phytohormone (15-carbon compound that resembles the terminal portion of some carotenoid molecules) involved in stress responses; its name is derived from its involvement in leaf abscission; it reduces cell expansion and causes stomatal closure.
- Substance that promotes the abscission.
- Plant growth hormone accumulates in leaf tissues during stress causing stomatal closure.
- Substance accumulates during drought conditions, causes stomatal closure, inhibits growth, promotes leaf, fruit and flower inhibition and induces dormancy in seeds of many species.
- An endogenous growth hormone involved in dormancy induction, seed maturation and induction of desiccation tolerance.

Abscisic acid insensetives (AB13) and FUSCA3 : Genes necessary for the initiation of dormancy.

Abscission :

- A process whereby specific cells the petiole differentiate to form an abscission layer, allowing the senescent organ to seperate from the parent plant.
- Detachment of leaves, flowers and fruits from plant caused by the formation of abscission layer.
- Shedding of organs such as leaves, flower, fruits in response to the formation of abscission layer which secretes hydrolytic

enzymes cellulase and pectinase leads to dissolution of cell wall and organ detaches at this point.

- The shedding of leaves, flowers, and fruits from a living plant. The process whereby specific cells in the leaf petiole (stalk) differentiate to form an abscission layer, allowing a dying/dead organ to separate from the plant.

Abscission layer :

- A layer of cells formed across the vascular conducting tissue in peduncle / funicle that aid shedding of seed/leaf from the mother plant. **Syn.** Abscission zone
- Distinct layer of cells within the abscission zone.
- Located within the abscission zone, a distinct layer of cells with weakened cell walls that permit abscission, usually of a leaf or fruit.

Abscission Zone :

- A region that contains the abscission layer and is located near the base of the petiole of leaves.
- Plant parts viz. leaves, flower, fruits abscise in a region.

Absolute humidity :

- Amount of total moisture in atmosphere.
- Amount of water vapour actually in the air, expressed either in its expansive force or in its weight per given volume, as gains per cubic foot.

Absorption :

- When a molecule, ion or atom enters inside the cell.
- Imbibing of water by living cells or tissues in a, seed.

Absorption index : An index value that indicates the ability of the genotype to imbibe water. Lower the index, higher the percentage of hard seeds.

AI = A/B x 100/y

Where A = weight of soaked seeds after 24 hrs;

B = initial weight of seeds;

Y = Number of hard seeds.

Absorption spectrum : A graphic representation of the amount of light energy absorbed by a substance plotted against the wavelength of the light.

ACC : The immediate precursor of ethylene; 1- aminocyclopropane - 1-carboxylic acid.

ACC (1 - amino cyclopropane - 1 - carboxylic acid) : An ethylene precursor.

ACC (1 - amino cyclopropane - 1 - carboxylic acid) synthase : The enzyme that catalysis the conversion of Adomet to ACC.

ACC oxidase :

- Catalyzes the conversion of ACC to ethylene, the last step in ethylene biosynthesis.
- Enzyme catalysis the last step in ethylene biosynthesis i.e. conversion of ACC to ethylene.

ACC synthase : The enzyme that catalyzes the synthesis of ACC from S-adenosylemethionine.

Accelerated ageing : A vigour test to assess the storage potential of seeds by subjecting the seeds to high humidity. (80% - 100%) and temperature (40°C) for specific period. The germination percent after exposure would indicate seed storability.

Accelerated ageing test :

- Exposure of seeds to 40-45° C at 100% humidity for 2-8 days and later subjecting seeds to regular germination test.
- Intentionally subjecting seeds to adverse storage conditions for a short period to estimate their possible life span under favourable storage conditions.

Acclimation (hardening) :

- The increase in plant stress tolerance due to exposure to prior stress. May involve gene expression. Contrast with adaptation.
- Increased tolerance to stress and/or improved plant performance as a result of structural and physiological adjustment by individual plants to specific environmental conditions.

Accumulation : Build-up of storage products resulting from an excess of supply over demand; also termed interim deposition.

Acetogenic bacteria : Obligatory anaerobes that use CO_2 and H_2 as electron acceptor and donor, respectively, in anaerobic respiration for the production of acetyl-CoA. The latter compound serves as building blocks for biosynthetic processes while the excess is excreted as acetate.

Acetyl COA : Synthesis is the link between kreb's cycle and glycolysis.

Achene : A single seeded dry, indehiscent fruit of *Astraceae (Compositae)* family.

Acid delinting : A process of removal of fuzzy from cotton seed using concentrated Sulphuric Acid.

Acid growth : A characteristic of growing cell walls in which they extend more rapidly at acidic pH than at neutral pH.

Acid-growth hypothesis : Stress relaxation and expansion of primary cell walls that is induced by protons transported across the plasma membrane.

Acidifuge : Avoiding acid soils, with a preference for a substrate that does not have a low pH Active (or reactive) oxygen species (ROS) hydrogen peroxide (H_2O_2), superoxide radicals (O_2.–), and hydroxyl radicals (OH.), the compounds can cause cell damage, but are also involved in signal transduction.

Acid treatment : Refers to acid scarification of seeds (to overcome hard seed coat). Seeds are placed in commercial Sulphuric Acid (1/2 volume of seeds), stirred for 5-20 minutes depending upon species and washed to remove aid. This treatment improves germination.

Acropetal : From the base to the tip of an organ, such as a stem, root, or leaf.

Action spectrum : A graphic representation of the magnitude of a biological response to light as a function of wavelength.

Activated alumina : Highly porous, granular form of aluminum oxide, with high adsorptive capacity for moisture.

Active absorption : Absorption process which involves metabolic energy. The movement of substances may occur against or upgradient or chemical potential.

Active ingredient : Refers to the chemical present in the herbicide / pesticide that is toxic to certain / all plants/ pests, **Syn.** AI

Active transport :

- The use of energy to move a solute across a membrane against a concentration gradient, a potential gradient, or both (electrochemical potential). Uphill transport.
- Transport of molecules across a membrane against an electrochemical gradient through expenditure of metabolic energy.

Acyanogenic : Not releasing cyanide.

Acyl : Refers to the fatty acid residue linked to another compound, often glycerol (e.g., triacylglycerol).

Acyl-ACP : A fatty acid chain bonded to the acyl carrier protein.

Acyl carrier protein (ACP) :

- A low molecular weight, acidic protein to which growing acyl chains are covalently bonded on fatty acid synthetase.
- Acyl chains bound to a low molecular weight acid protein.

Adaptation : Evolutionary adjustment of the genetic basis of a trait that enhances the performance in a specific environment.

Adaptation (to stress) : An inherited level of stress resistance acquired by a process of selection over many generations. Contrast with acclimation.

Adaxial :

- Located on the side nearest to the axis, e.g., the upper side of a leaf.
- Refers to the upper surface of a leaf.

Adenosine-5′-phosphosulphate (APS) : A short- lived, activated form of sulfate, formed by the reaction between sulphate and ATP. Product of first reaction in the several pathways from sulphate to the amino acid cysteine.

Adequate zone : A point at which further addition of nutrients is no longer related to increases in growth or yield.

Adhesion : Attraction of solvent to a solid phase such as cell wall or glass surface.

Adjuvant (herbicide) : A chemical additive to the herbicide that improves the action of herbicide.

AdoMet : S-adenosylmethionine.

ADP/ATP transporter : A protein that catalyzes an antiport exchange of ADP and ATP across the inner mitochondrial membrane.

ADP/O ratio : The ratio of consumed ADP to $^{1/2}O_2$ in oxidative phosphorylation. Provides the number of ATP synthesized per two electrons transferred to oxygen.

Adsorption :

- Binding of ions or molecules to a surface (e.g., of a soil particle or a root).
- Taking up a gas, vapour, or dissolved material on the surface of a solid; finely divided materials, as active carbon and silica gel, can adsorb relatively large quantities of other materials.
- When a molecule, ion or atom adhere over some surface.

Advection : Net horizontal transfer of gases. Aerenchyma : Tissue with large air spaces that facilitate transport of gases in plants.

Adventitious : Development of organs (shoots, roots, buds, flowers etc.) i.e. sterile.

Adventitious embryo :

- Bipolar body that develops in to a plantlet with no need for any seperate rooting but embryo lacks food reserves.
- A condition in seed where the embryo arises from somatic tissue. e.g. Citrus. Syn. Nucellar embryony

Adventitious roots :

- A root arising from the mesocotyl region of the embryo.
- Roots that arise from structures other than roots, such as stems or leaves.

Aerenchyma : Anatomical feature of roots found in hypoxic conditions, showing large, gas-filled intercellular spaces in the root cortex.

Aerobic respiration :

- Respiration in which O_2 is required.
- Biological process by which reduced organic compounds are metabolized and subsequently oxidised in a controlled manner. Free energy is released and stored in ATP.

  $C_6H_{12}O_6 + 6O_2 + 6\ H_20 \xrightarrow[\text{Chlorophy}]{\text{Light}} 6\ CO_2 + 12\ H_20 + 673$ K.cal energy.

  Or $C_6H_{12}O_6 + 6O_2 + 6\ H_20 + 38\ ip + 38\ ADP \xrightarrow[\text{Chlorophy}]{\text{Light}} 6\ CO_2 + 12\ H_20 + 38\ ATP$

Aeroponics : System in which roots are suspended over the nutrient solution, which is whipped into a mist by a motor driven rotor.

After ripening :

- A large number of seeds do not germinate immediately after harvesting but if kept for some period of time under favourable conditions they start germinating.
- Many seeds lose dormancy when their moisture content is reduced to a certain level by drying.
- The process in which the seed with immature embryo is made to mature by placing it under warm-moist condition.

Agar block : A semi-solid cube of agar used to donate or receive auxin when in contact with plant tissue. Used in studies of polar auxin transport.

Agar culture : Allowing the protoplasts to regenerate cell wall in liquid culture before they are mixed with .6% agar or placed in Agar culture media as soon as they are isolated.

Agarose : A linear polymer of carbohydrate extracted from seaweed.

Agarose gel electrophoresis : Standard method for the separation, identification and purification of DNA and RNA fragments, ranging in size from a few hundred to 20 kb.

Age index : A method to judge the physiological age of stored seed propounded by Húkill (1963).

Ageing test : Any test to determine the deterioration of seeds with their increased age, usually a germination or vigor test.

Agglutinin : Synonym for lectin.

Agravitropic mutants : Mutants that do not respond to gravity. Useful in understanding the mechanism of gravitropism.

Agrobacterium : Transformation vector used to deliver foreign DNA (not normally present in recipient genome) in to plants.

AHPs genes : Arabidopsis Histidine Phosphotransfer genes involved in cytokinin signal propagation from the plasma membrane receptor to the nucleus.

Air oven method :

- Method of moisture estimation by applying heat to expel the moisture present in seeds using hot air oven. This is the only method recognized by ISTA.
- Drying seeds in a forced air oven to determine their moisture content.

Albedo : Fraction of the incident short-wave radiation reflected by a surface (typically plant cover or bare soil or rock).

Albuminous seed : Seeds with endosperm. e.g. Maize, rice, castor etc.

Alcohol dehydrogenase (ADH) : The enzyme that catalyzes the conversion of acetaldehyde to ethanol.

Alcoholic fermentation : Metabolism of pyruvate from glycolysis produces ethanol and CO_2, while reoxidizing NADH to NAD^+.

Aleurone cells : Cells of the aleurone layer enclosed in thick primary cell walls and containing large numbers of protein-storing organelles called protein bodies.

Aleurone layer :

- An outermost layer of living cells covering the endosperm that contains Alpha Amylase enzyme in pockets. It stains red during tetrazolium.
- Layer in monocot seeds like maize containing protein granules.

- Layer of aleurone cells surrounding and distinct from starchy endosperm of cereal grains.

Aleurone tetrazoliam vigour test : A tetrazoliam vigour test in maize. Presoaked maize seeds are given cuts on pericarp to expose the aleurone layer. Seeds are soaked for 2-3 days in Tz solution. The aleurone layer stains red and assessed for vigour based on the staining pattern.

***alf* mutants** : A series of Arabidopsis mutants (aberrant lateral root formation mutants) in which the role of auxin in the initiation of lateral roots is altered.

Alkaloids :

- Secondary plant compound (often toxic), characterized by its alkaline reaction and a heterocyclic ring (e.g., nicotine, caffeine, and colchicine).
- Large family of more than 15000 nitrogen containing secondary metabolites found in approximately 20% of the species of vascular plants but best known for their striking pharmacological effects on vertebrate animals. Usually synthesized from one of a few common amino acids, in perticular lysine, tyrosine and tryptophan.
- A large family of nitrogen containing secondary metabolites found in many vascular plants. Defend against predators, especially mammals. Includes toxins such as strychnine and atropine, and medicinals such as morphine, codeine, atropine, and ephedrine. Others are stimulants and sedatives (e.g., cocaine, nicotine and caffeine).

Allelochemicals :

- Secondary metabolites, released by living plants or decomposing plant litter that (either negatively or positively) affects other Organisms.
- Chemicals found in plants that induce allelopathy.

Allelopathy :

- It is effect of one plant affecting another plant's growth either positively or negatively by exuding characterized allelopathic chemicals on the germination, growth or development.

- Release by plants of substances into the environment that have harmful effects on neighboring plants.
- Suppression of growth of one plant by another of a different species due to the release of toxic substances.

Allocation :

- Proportional distribution of products or newly acquired resources among different organs or functions in a plant.
- The regulated diversion of photosynthate into storage, utilization, and transport.
- Allows the TCA cycle to function in the absence of oxygen.

Alpha amylase :

- An enzyme occurring widely in seeds, leaves, and other plant parts; it accelerates starch hydrolysis.
- Enzyme involved in starch degradation during early germination.

Alpha (α) Oxidation : Major pathway for degradation of fatty acids especially in animals in which á carbon (carbon - 2) is oxidised to form new COOH group.

Altenative oxidase :

- A plant and fungal specific enzyme in the mitochondrial electron transport chain, which reduces oxygen to water.
- Mitochondrial enzyme catalyzing the transfer of electrons from ubiquinol (the reduced form of ubiquinone) to O_2.

Alternating temperature : A method of maintaining temperature in germination room where the seeds placed for germination are exposed to higher (16 hrs.) and lower temperatures (8 hrs.) for fixed duration during germination period.

Alternative pathway (of respiration) :

- Nonphosphorylating electron-transport pathway in the inner membrane of plant mitochondria, transporting electrons from ubiquinol (the reduced form of ubiquinone) to O_2, catalyzed by the alternative oxidase.
- The pathway comprising the oxidation of ubiquinol and the reduction of oxygen by the alternative oxidase. Also known as cyanide-resistant pathway.

Ambient : That which encompasses or surrounds on all sides a designated object, state of matter, system, etc.; as used here, ambient refers to temperature or relative humidity of the atmosphere in the natural state without modification by man.

Ambiphotoperiodic : Describes plant that flower in long days or short days but not at intermediate day lengths.

Amide : A nitrogen-containing compound formed by the reaction of an amine and a carboxylic acid to form a $-CONR_2$ group.

Amide exporters : Legumes from temperate region that convert toxic ammonia to amides, such as the amino acids asparagine or glutamine for transport to the shoot via the xylem. Contrast with ureide exporters.

Amine : A nitrogen-containing compound derived from ammonia by replacing hydrogens with carbon-containing groups.

Aminoethoxy - venyl - glycine (AVG) and amino oxyacetic acid (AOA) : Inhibitors of ethylene synthesis which block the conversion of Adomet (S adenosyl methionine) to ACC, AVG and AOA are known to inhibit enzymes that use the cofactor pyridoxal phosphate.

Aminotransferases : Enzymes that carryout transaminations.

Ammonical nitrogen : Reduced form of nitrate i.e. NH_3 absorbed by some plants directly like paddy. Nitrogen in ammonical form is incorporated in to nitrogenous compounds of the plant.

Ammonification : Conversion of amino nitrogen to ammonia by soil bacteria and fungi.

Ammonium chloride soak test : A simple vigour test to differentiate vigourous and non-vigourous seed lots. The seeds are soaked in 4% ammonium chloride for 16 hrs. and then germinated to assess the vigour.

Amphipathic : In a molecule, the quality of having both hydrophilic and hydrophobic regions.

Amphistomatous with : Stomata at both the adaxial (upper) and abaxial (lower) sides of a leaf.

Amplitude :

- Distance between peak and trough. The amplitude of a biological rhythm can often vary while the period remains unchanged.
- In a biological rhythm, the distance between peak and trough; it can often vary while the period remains unchanged.

Amylase : Enzyme associated with starch breakdown.

Amylopectin : Has a side chain of glucose with B 1-6 linkage.

Amyloplast : A starch storing plastid.

Amylose : Component of starch consists of 300-400 glucose molecules.

Anabolic process : Process by which matter is changed into tissues of living plants or seeds.

Anaerobic respiration : Respiration process in which oxygen is not required.

$C_6H_{12}O_6 \rightarrow 2\ C_2H_5OH + 2\ CO_2 + 54$ K.cal. Energy

Anaplerotic : Quality of a reaction to provide a reactant that is limiting in another reaction or pathway for example, the PEP carboxylase reaction replenishes oxaloacetate to the citric acid cycle that has been used for biosynthesis.

Anastomoses (Singular Anastomosis) :

- In absence of direct connections between source and sink, vascular interconnections.
- Vascular interconnections that provide a pathway between source and sink tissues that are not directly connected.

Anchored proteins : Proteins bound to the membrane surface via lipid molecules to which they are covalently attached.

Ancymidol (aka : A- Rest) : A commercial inhibitor of gibberellin biosynthesis. Like paclobutrazol, ancymidol blocks the oxidation reaction between kaurene and kaurenoic acid on the endoplasmic reticulum.

Angiosperm : Flowering plants, distinguished from gymnosperms by the presence of a carpel that encloses the seeds.

Anion : Negatively charged ion.

Anisotropic :

- Characteristic of materials showing different mechanical properties in different directions. Usually due to a bias in the orientation or linkage of constituent molecules. Contrast with isotropic.
- Not equal in all directions; for example, the longitudinal walls of anisotropic cells have different chemical and biophysical properties from those of the radial walls.

Annealing : Cooling of mixture at 40-60^0 C that permits annealing of the primer to the complementary sequences in the DNA located at the 3 ends of the two strands of the desired segment.

Annual species : With a life cycle of less then a year; the short life cycle can be environmentally or developmentally determined.

Anoxia : Absence of oxygen in (part of) a plant's Environment.

Anoxic : Refers to the absence of oxygen. Contrast with hypoxic.

Anoxygenic organism : Photosynthetic organism that does not produce molecular oxygen.

Antenna complex : A group of pigment molecules that cooperate to absorb light energy and transfer it to a reaction center complex.

Antenna pigments : Chlorophylls and carotenoids are the pigments found in the antenna complex.

Anther :

- Male sex organ containning pollen grains.
- The apical structure of stamen in which pollen forms and from which it is released.

Anther culture : Culture of anthers and or immature pollen grains to obtain haploid cells.

Anthocyanidins : Pigmented flavonoids derived from anthocyanin by removal of attached sugars.

Anthocyanins :

- Widespread group of pigmented flavonoid responsible for most of the red, pink, purple and blue colours observed in plant parts.

These are glycosides that have sugars at the position three and sometimes elsewhere.

- Pigmented glycosylated flavonoids responsible for most of the red, pink, purple, and blue colors in plants.

Anticlinal cell division :

- Transverse cell division with the plane of cytokinesis oriented at right angles to the axis of root (such divisions tend to increase root length). Examples - Cell division in roots.
- New wall formed perpendicular to the closest surface.
- Pertaining to the orientation of the cell plate at right angles to the longitudinal axis during cytokinesis. Transverse.

Antiflorigen :

- Hypothetical hormone produced by uniduced leaves and translocated to the shoot apical meristem. Proposed to inhibit the formation of flowers in certain long-day plants under noninductive conditions.
- Transmissible inhibitors in flowering regulation.

Antifreeze proteins :

- Proteins that confer to aqueous solutions the property of thermal hysteresis. When induced by cold temperatures, these plant proteins bind to the surfaces of ice crystals to prevent or slow further crystal growth, thereby limiting or preventing freeze damage. Some antifreeze proteins may be identical to pathogenesis-related proteins.
- Proteins which have the ability to inhibit ice crystal growth in a noncolligative manner, thus preventing freeze damage at intermediate freezing temperatures.

Antimycin a : A specific inhibitor of complex III of the election transport chain of the mitochondrion.

Antioxidant synergists : Chemicals added along with antioxidants to improve their performance. e.g. Phosphates of sodium, Potassium etc.

Antioxidants : A group of chemicals that react with the free radicals present in membrane systems and prevent free radical damage. e.g. Vit. E, C, Beta Carotene etc.

Antipodal cells : Group of three cells located at the upper side of embryo sac.

Antipodal nuclei : Three of eight nuclei in the ovule that are found opposite to the egg cell.

Antiport :

- A type of secondary active transport in which the passive (downhill) movement of protons of other ions drives the active (uphill) transport of a solute in the opposite direction.
- Co-transport of one compound in one direction coupled to transport of another compound (mostly H+) in the opposite direction.
- Coupled transport in which the downhill movement of proteins drives the active (uphill) transport of a solute in the opposite direction.

Antiporter : A protein involved in antiport.

Antisense DNA : DNA of a gene whose transcription produces anti-sense mRNA, which hybridizes with sense mRNA and inhibits its translation.

Antitranspirants : Substances which reduce transpiration rate by causing stomatal closure partially. Examples-colourless plastics, silicone oil, low viscocity waxes, phenyl mercuric acetate, abscisic acid, kaolin, CO_2 above .03%.

AOA : Aminooxyacetic acid, an inhibitor of ethylene biosynthesis, blocking the conversion of *AdoMet* to ACC.

AOSA : Acronym for Association of Official Seed Analysts; the organization of State and Federal seed analysts of USA and Canada.

AOSCA : Acronym for association for Official seed certifying agencies; the organization of certification agencies of USA and Canada.

AP1 gene : In *Arabidopsis,* APETALA1 (AP1) is a gene involved in establishing floral meristem identity.

AP2 mutants : In Arabidopsis, mutations in APETALA2 (AP2) homeotic gene produce flowers with carpels where speals should be and stamens where petals normally appear.

Apatite $Ca_5 (PO_4)_3(OH,F)$: Accounts for 95% of the total P in igneous rock, and it constitutes a major substrate for weathering, which releases inorganic phosphate for plants and microorganisms.

Apical : Relating to the apex or tip. Distinguishing one end of an axis. Contrast with *basal.*

Apical cell : In ferns and other primitive vascular plants, the single initial of stem cell of roots and shoots that gives rise to all the other cells of the organ. In angiosperm embryogenesis, the smaller, cytoplasm-rich cell formed by the first division of the zygote.

Apical dominance :

- Growth in vertical direction through suppression of lateral growth.
- In most higher plants, the growing apical bud's inhibition of the growth of lateral buds (axillary buds).
- Inhibition of growth of apical bud due to growth of lateral (axillary) buds.
- The phenomenon of inhibition of the development of lateral buds by the growth of apical meristem.

Apical meristems : Located at top of stem and root.

Apoenzyme : Enzymatic protein that requires a coenzyme to function.

Apomixis :

- Formation of embryo by the process other than union of male and female gametes.
- Seed development without sexual fusion of the egg and sperm or development of embryo without fertilization.

Apoplast (¼apoplasm) :

- Space in a plant's tissue outside the space enclosed by plasma membranes.
- Refers to the dead parts of plant cells viz. all walls, xylem vessels, intercellular spaces.
- The mostly continuous system of cell walls, intercellular air spaces, and xylem vessels in a plant.

Apoplastic (¼apoplasmic) : Phloem loading transport of assimilates from mesophyll to the sieve tubes of the phloem occurring partly through the apoplast.

Apoplastic pathway : The route by which water and water-soluble solutes move exclusively through the cell walls without crossing any membranes.

Apoprotein :

- A polypeptide chain, one of the components of phytochrome.
- The polypeptide component which may be bound to a chromophore or other cofactor or prosthetic group to form an active protein, the holoprotein.

Apoptosis :

- Distinct set of morphological and biochemical changes in which the cell nucleus condenses and the nuclear DNA fragments in a specific pattern caused by the degradation of the DNA between nucleosomes.
- Programmed cell death showing characteristic morphological and biochemical changes, including fragmentation of nuclear DNA between the nucleosomes. Occurs in some senescing plant tissues, differentiating xylem tracheary elements, and in the hypersensitive response against pathogens.

Apparant free space : Apparant volume of plant tissue for accomodating the diffusion of ions freely.

Apple and Mulberry type : Stomata on upper leaf surface. Examples - Peach, Walnut, Nasturtium, Oxalis, Yew etc.

Apposition : Layering of cellulose polymer over cellulose.

Aquaporin :

- Water-channel protein in a membrane.
- Integral membrane proteins that form water selective channels across the membrane.
- A class of proteins relatively abundant in plant membranes facilitates rapid influx of water across oocyte plasma membrane, which normally has a very low water permeability.

- Integral membrane proteins that form water-selective channels across the membrane. Such channels facilitate water movement across the membrane.

Arabidopsis Response Regulator : Arabidopsis genes that are similar to bacterial two-component signaling proteins called response regulators. There are two classes : Type-A ARRs, whose transcription is up-regulated by cytokinin; type-B ARR, whose expression is not affected by cytokinin.

Arabinogalactan proteins (AGPs) :

- Water soluble, highly glycosylated cell-well proteins that may function in cell adhesion and signaling during cell differentiation.
- A class of glycoprotein macromolecules that may communicate positional information that will determine cell fate.

Arbuscular mycorrhiza : A type of mycorrhiza that forms arbuscules (highly branched exchange structures) within cortical cells of the root.

Arbuscules : Branched structures of mycorrhizal fungi that form with in penetrated cells; the sites of nutrient transfer between the fungus and the host plant.

Arenchyma - Separation of cells by prominent gas filled spaces.

Arnon and Stout (1939) : Element must be essential for growth and reproduction of plant with specific requirements and direct involvement in plant system.

Artificial ripening : A technique used for ripening pine cones collected prior to maturity. The cones are placed in ventilated cellars using forced air circulation at species specific temperatures for 2-3 months.

Artificial seeds :

- Somatic embryos enclosed in calcium alginate gel used for propagation.
- Encapsulated embryo produced by somatic embryogenesis.

Ascent of sap : Upward movement of water.

Asceptic : Free from all microorganisms.

Asexual reproduction : A type of reproduction that does not involve the fusion of male and female gametes.

Asparagine synthetase (AS) : Enzyme transferring nitrogen as an amino group from glutamine to aspartate, forming asparagine.

Aspartate aminotransferase (Asp AT) : An aminotransferase that transfers the amino group from glutamate to the carboxyl atom of oxaloacetate to form asparatate.

Aspirator : An air-blast separator; a pre-cleaning equipment to separate seeds from thrash and other light weight material by blowing away the light weight materials.

Assimilation : Incorporation of an inorganic resource (e.g., CO_2 or NH4+) into organic compounds (in the case of CO_2 assimilation also used as a synonym for photosynthesis).

Assimilatory cotyledons : Thin green leaf like cotyledons that provide the necessary assimilates upon germination till true leaves develop. e.g. cotton, castor.

Assimilatory power : The combined energy available in NADPH and ATP that can be used to drive the photosynthetic flxation of atmospheric CO_2 into organic molecules.

Asymbiotic N fixation : Fixation carried out by free living organisms. Examples - Anaerobic bacteria - Clostridium, Aerobic bacteria - Azotobacter and blue green algae - Anabaena.

AtNHX1 gene : In *Arabidopsis,* gene encoding a Na^+/H^+ antiporter that is active in vacuolar compartmentalization of Na^+.

ATP : Acronym for Adenosine Triphosphate, the source of biological energy in the cells.

ATPase : Enzyme catalyzing the hydrolysis of ATP, producing ADP and Pi; the energy from this hydrolysis is used to pump protons across a membrane (e.g., plasma membrane, tonoplast), thus generating an electrochemical gradient.

ATP synthase (ATPase or CF_0-CF_1) :

- The enzyme that synthesizes ATP from ADP and phosphate (P). Consists of two parts : a hydrophobic membrane bound portion (CF_0) and a portion that sticks out into the stroma (CF_1).

- Enzyme complex in the inner membrane of mitochondria and the thylakoid membrane of chloroplasts catalyzing the formation of ATP, driven by the proton-motive force (pmf).

ATP-binding cassette transporters (ABC transporters) : A group of active transport proteins, of distinctive structure, energized by ATP hydrolysis and involved in moving organic molecules across a membrane.

Autamutagens : Mutagens which bring about mutation.

Autocatalysis : The capacity of the Calvin cycle to use most of the fixed carbon (triose phosphate) to build up the concentrations of intermediates to adequately function at steady state. As a result, the concentration of the acceptor (ribulose-1,5 bisphosphate) doubles every five revolutions of the cycle.

Autocatalytic :

- An action or reaction that promotes the action or reaction. For example, ethylene production by fruit is stimulated by ethylene.
- Production of additional ethylene by fruit itself in response to ethylene treatment for attaining ripening.
- Relating to a process that occurs spontaneously when purified components of a system are mixed in vitro, requiring no additional proteins or cofactors.

Autogamy : Fertilization of ovum by the pollen of same flower. Syn. Selffertilization.

Autoinhibitory domain : A domain at the C-terminal end of the plasma membrane H^+-ATPase that inhibits the activity of the enzyme. Regulated by phosphorylation and by agents such as fusicoccin.

Auto-oxidation : Refers to the oxidation of lipids by breaking of hydroxyl radical that inturn affects other lipids thereby starting a chain reaction in biomembranes. It is mainly caused due to ageing, higher temperature, moisture etc.

Autoradiography : A technique used to determine the location of a radioactive isotope that has been supplied to living cells, using photographic film (overall localization) or photographic emulsion (precise cellular localization).

Autotoxicity : Deleterious effect of a chemical compound released by plants of the same species.

Autotrophic : Cells or organisms that synthesize cellular components from structural elements found in the most oxidized states (for example, carbohydrates from CO_2). As the reduction is usually endothermic the energy is supplied either by sunlight (photoautotrophs) or by chemical energy (chemoautotrophs).

Autotrophic growth : Increment in mass, volume, length, or area of plants or parts thereof which depend on carbon fixed in photosynthesis by the growing organism itself (see also heterotrophic growth).

Autotrophic respiration : Respiration by autotrophic plants and their associated mycorrhizas and symbiotic N2-fixing structures (see also heterotrophic respiration).

Aux/IAA genes : Family of primary response genes encoding short lived transcription factors that function as repressors or activators of the expression of late auxin inducible genes.

Auxamometer : Instrument used to record plant growth.

Auxin :

- A compound with biological activities similar to, but not necessarily identical with, those of IAA. Induces cell elongation in isolated coleoptile or stem sections, cell division in callus tissues in the presence of a cytokinin, lateral root formation at the cut surfaces of stems, parthenocarpic fruit growth, and ethylene formation.
- Phytohormone (indole-3-acetic acid) involved in growth promotion and meristem differentiation ; the name literally means enhancing and is derived from its growth-promoting action; there are also synthetic auxins.
- A group of growth regulating substances that stimulate cell growth, root development and seed germination.
- Substances useful in stem growth, cell elongation, secondary growth, rooting in cutlings, apical dominance, production of female flowers, parthenocarpic fruit development, prevention of leaf and fruit abscission and phototropism.

Auxin-Induced Proton Extrusion : Increased rate of proton extrusion by both activation of preexisting plasma membrane H^+-ATPases and synthesis of new H^+-ATPases.

Auxin response domains (AμxRDs) :

- Auxin responsive promoter sequences composed of multiple AuxREs.
- Union of a conserved auxin response element (Aux RE) within the promoters of the early auxin genes like GH3 with other response elements.

Auxin response elements (AuxREs) : DNA promoter sequences that modulate gene expression when bound by auxin-responsive transcription factors.

Auxin response factor (ARF) : Protein related to transcription factor encoded by MP gene.

Auxin transport : The energy requiring, polar movement of IAA from the shoot apex to the root tip, and subsequent redistribution of auxin from the root tip to the basal portion of the root. Consists of protein mediated auxin efflux from cells driven by the membrane potential, ÄE, and auxin uptake into cells driven by membrane potential, ?E and auxin uptake into cells driven by the total proton motive force (ÄE+ÄpH).

Auxin transport inhibitor (ATI) : Any compound that prevents the polar transport of auxin, usually by preventing auxin exit (efflux) from cells.

AVG : Aminoethoxy vinylglycine, an inhibitor of ethylene biosynthesis, blocking the conversion of AdoMet to ACC.

Avoidance :

- Plant strategy of resisting adverse conditions by preventing deleterious effects of these conditions, e.g., winter seed dormancy.
- Some plants avoid stress like in desert a plant avoid dry soils by extending its roots down to water table.

Avr genes : Pathogen avirulence genes that encode for specific elicitors of plant defense responses.

Axil : The angled juncture between the upper side of a leaf and the stem to which it is attached. Usual site for insertion of an axillary bud.

Axis (plant axis) : The hypothetical central line of a body around which parts are arranged.

Axillary : Originating in the axils of the leaves.

Axillary buds :

- Bud developed in the angle between leaf and stem.
- Secondary meristems that are formed in the axils of leaves. If they are also vegetative meristems, they will have a structure and developmental potential similar to that of vegetative apical meristem. Axillary buds can also form flowers, as in inflorescences.
- Secondary meristems, if they are also vegetative meristems, they will have a structure and developmental potential similar to that of apical meristem.

Axillary meristem : Meristems formed in the axils of leaves dervied from the shoot apical meristem.

Axil mutant : Arabidopsis auxin resistant mutant deficient in many auxin response deficiencies, including gravitropism.

Axil patterning : Axil polarity that almost all plant possess in which the tissues and organs are arrayed in a precise order along a linear or polarized axis. Shoot apical meristem is at one end of the axis and root apical meristerm at the other.

□□□

B

Bacterial Photosynthesis : Manufacture of food material using light energy, H_2S or other compound for electron source utilizing pigments in small structures designated as chromatophores. No O_2 evolution. Examples chlorobacterium, chromatium, Rhodospirillum.

Bacteriochlorophyll : Light absorbing pigments active in photosynthesis in anoxygenic photosynthetic organisms.

Bacteroid :

- Conversion of bacterium infection thread into globular oval cells.
- State of rhizobia after they have penetrated the root and the symbiosis has been established.
- Nitrogen-fixing organelles that develop from endosymbiotic bacteria upon a signal from the host plant.

Bar and Megapascal (MPa) : Units to express water potential.

Bark :

- Collective term for all the tissues outside the cambium of a woody stem or root, and composed of phloem and periderm.
- Tissue with both a protective (outer bark) and transport (inner bark) function; inner bark consists of secondary phloem that carries sugars, amino acids, and minerals from a source to a sink.

Barnase gene : Prevents normal pollen development by killing the tapetal cells leading to male sterility.

Basal : Relating to the base. Distinguishing one end of an axis. Contrast with apical.

Basal cell : In embryogenesis, the larger, vacuolated cell formed by the first division of the zygote.

Basic seed : In Indian seed certification context refers to the seed produced by mass selection in a pure line variety or clone used for production for Nucleus seed.

Basipetal : From the growing tip of a shoot or root toward the base (junction of the root and shoot).

Basipetal transport : Transport away from the apical meristems, in both root and shoot.

Batch cultures : Cultures initiated as single cells in 100-250 ml flasks and propagated by transferring regularly small aliquots of suspension to a fresh medium.

Beta-(β-)oxidation :

- Oxidation of fatty acids into fatty acyl-C_OA, and the sequential breakdown of the fatty acids into a number of pairs of two molecules of acetyl-C_OA. NADH and $FADH_2$ are also produced.
- Oxidation of functional group present in the B position of carbon atom (i.e. carbon three from the COOH end) to produce new COOH group.

Between Paper method (BP) : A method of placing seeds for germination, where the seeds are placed between two layers of moist germination paper.

Bidirectional transport : Simultaneous transport in two directions in a single sieve element.

Biennial : Plant that requires two growing seasons to flower and produce seed.

Biennial species : Whose individuals typically live for two growing seasons, vegetative growth in the first year and continued growth and seed production in the second year; several species known as biennials can, however, have an extended vegetative period (monocarpic perennial), others are strictly biennial.

Bioassay : Quantitation of a known or suspected biologically active substance by measuring the effect the substance has on living material.

Biochemical change : Change in the chemistry of a plant, plant part, or seed.

Biological nitrogen fixation :

- Certain organisms utilize molecular nitrogen of atmosphere and synthesize nitrogenous compounds.
- Nitrogen fixation carried out by bacteria or blue-green algae (cyanobacteria); about 90% of all nitrogen fixation on Earth.

Biomass :

- Dry weight of whole plant including roots, also expressed as organic matter or energy content of plant.
- Mass of plants (and other living organisms).

Biomass density : Dry mass of plant tissue per unit of fresh mass or volume (in the first case, the presence of intercellular air spaces is not taken into account).

Biomass duration (BMD) : Parameter represents dry weight losses or gains –

Kvet (1962) $BMD = \frac{W_2 - W_1 (t_2 - t_1)}{Ln\, W_2 - Ln\, W_1}$ Weight Time

Bioprocesses : Use of living cells or tissues, their components (e.g. enzymes and chloroplasts) in a controlled and directed manner so as to produce some desirable physical or chemical changes.

Bioreactors : Advanced methods for conducting large scale bioconversions with cells or their components.

Biotic :

- Caused or induced by organisms.
- Referring to what is living.

Biotic filter : Biotic interactions which eliminate species that would otherwise have survived the abiotic environment of a site.

Bio-transformation : Transformation with the help of microbes to produce valuable products.

Bleached : The loss of chlorophyll's characteristic absorbance due to its conversion into another structural state, often by oxidation.

Bleeding : Exudation of liquid water, sap and dissolved substances from injured parts of plants.

Blue light responses : Responses of plant cells and organs to blue light (400to500nm). Includes phototropism, chloroplast movement within cells, sun tracking by leaves, inhibition of hypocotyl elongation, stimulation of chlorophyll and carotenoid synthesis, activation of gene expression and stomatal movements.

Blue-light receptors : Two classes of photoreceptors, cryptochromes and phototropins, that absorb in the blue region of the spectrum; the receptors are involved, e.g., in the perception of irradiance and the directional component of light and thus affect photomorphogenesis.

Boerner divider : A divider working based on gravitational force for mixing and dividing the seeds. It consists of an inverted cone whose periphery has baffies dividing the falling seed into two equal haives. Syn. Conical divider.

Bolting :

- Premature elongation of the stem of a rosette plant, usually associated with flowering.
- Rapid extension of the flowering stalk.

Boundary layer :

- A thin film of still air at the leaf surface. Its resistance to water vapour diffusion is proportional to its thickness.
- Thin layer of air, water, or soil around the leaf or root with reduced mass. Transport and increased reliance on diffusion for transport processes, conditions differ from those further away.

Boundary layer conductance/resistance :

- Conductance/ resistance for transport of CO_2, water vapour or heat between the leaf surface and the atmosphere measured across the boundary layer.
- The resistance to the diffusion of water vapour due to the layer of unstirred air next to the leaf surface. A component of diffusional resistance.

Bowen ratio :

- The ratio between sensible heat loss and heat loss due to transpiration.
- The ratio of sensible heat loss to evaporative heat loss, the two most important processes in the regulation of leaf temperature.

Bract : Small leaf-like structure with underdeveloped blade.

Branch root meristems : Meristems which have the structure of primary root meristem, but they form from pericycle cells in mature regions of the root.

Brick girt test : A vigour test that determines the ability of the seed to emerge through poor soil conditions. Seeds are covered with brick grit layer of 1cm. depth. Seeds that penetrate the layer and germinate are vigourous Syn. Brick grit test, Hiltner's test.

Brick gravel test : Germination test conducted by using 2 - 2.5 mm moist gravels placed above the seed.

Brownian movement : Zig zag movement of colloidal particles near the same spot in a suspension.

Bud culture : Culture of buds contain quiescent or active meristems.

Buffers : Solution of weak acid and strong salt or vice versa. Buffer resists any change in the pH of the solution. Example - Solution of acetic acid (weak) and NaCl (strong salt).

Bulk density :

- Mass of dry soil per unit volume.
- Weight of unit mass of seeds in relation to its volume. Expressed as kg/m^3.

Bulk flow (or mass flow) : The concerted movement of molecules in mass, most often in response to a pressure gradient.

Bulk soil : Soil beyond the immediate influence of plant roots.

Bundle sheath cells :

- Cells surrounding the vascular bundle of a leaf.
- Chloroplast containing cells found in the leaves of C_4 plants but not in C_3 plants.
- One or more layers of closely packed cells surrounding the veins of leaves and the primary vascular bundles of stems.

Bunning hypothesis : Hypothesis that photoperiodic control of flowering is achieved by a coincidence of the light or dark phase of an inductive photoperiod with a particular phase of oscillation of the circadian rhythm that has a different sensitivity to light.

❑❑❑

C

C - **Repeat binding factors (CBF1, CBF2, CBF3)** : Transcriptional activators of many cold stress - induced genes also called DREB1b, DREb1C, DREb1a, respectively (Shinozaki and yamaguchi - Shinozaki 2000).

C_3 **photosynthesis** : Photosynthetic pathway in which the first step of CO_2 assimilation is the carboxylation of ribulose 1,5-bisphosphate (RuBP) by Rubisco; the first product is phosphoglyceric acid (PGA), a three-carbon intermediate.

C_3 **plants** : Plants in which the first stable product of photosynthetic CO_2 fixation is a three carbon compound (i.e., 3-phosphoglycerate).

C_4 **cycle** : The photosynthetic carbon metabolism of certain plants in which the initial fixation of CO_2 and its subsequent reduction take place in different cells, the mesophyll and bundle sheath cells, respectively. The initial carboxylation is catalyzed by phosphoenolpyruvate carboxylase, (not by rubisco as in C_3

plants), producing a four-carbon compound (oxaloacetate), which is immediately converted to malate or aspartate.

C_4 photosynthesis : Photosynthetic pathway in which the first step of CO_2 assimilation is the carboxylation of phosphoenolpyruvate (PEP) by PEP carboxylase during the day; the first product is oxaloacetic acid (OAA) a four-carbon intermediate.

C_4 plants :

- First product of photosynthetic cycle is 4 carbon acid (Hatch and Slack, 1966). No differentiation of mesophyll cells into palisade and spongy parenchyma. Vascular bundles are surrounded with parenchymatus sheath which appears like wreath (Kranz anatomy). Examples – Sugarcane, Sorghum, maize, grasses, atriplex, panicum maximum, chloris guyana.
- Plants in which the first stable product of CO_2 assimilation in mesophyll cells is a four-carbon compound that is immediately transported to bundle sheath cells and decarboxylated. The CO_2 released enters the Calvin cycle.

Cadastral genes : Genes which act as spetial regulators of the floral organ identity genes by setting boundries for their expression.

Caking : Compaction of stored seed by high pressure caused by staking and high seed moisture.

Calcicole species : With a preference for calcareous or high-pH soils.

Calcifuge species : That typically occupies acidic soils and is absent from calcareous or high-pH soils.

Calcium – carbide method : A method of moisture estimation carried out by reacting water with C_aC_2 and measuring the acetylene released.

Callose : A β -1,3-glucan synthesized in the plasma membrane and deposited between the plasma membrane and the cell wall. Synthesized by sieve elements in response to damage, stress, or as part of a normal developmental process.

Callose b-(1-3) : Polymer of glucose, synthesized in sieve tube elements of the phloem in response to damage, sealing of the sieve tubes; callose is also produced in other cells upon microbial attack, thus providing a physical barrier.

Callus :

- A mass of undifferentiated cells produced in–vitro from various plant parts such as leaves and stem.
- Actively dividing non organized tissues of undifferentiated and differentiated cells often developing from injury (wounding) or in tissue culture.
- Undifferentiated group of cells.

Callus culture : Culture of a differentiated tissue from explant allowed to dedifferentiate in vitro.

Callus tissue : The product of the disorganized growth of undifferentiated plant cells in tissue culture.

Calmodulin :

- A conserved calcium binding protein found in all eukaryotes that regulates many calcium driven, metabolic reactions.
- A protein found in cytosol of plant cells. The calmodulin- calcium complex regulates many metabolic processes.
- Ubiquitous Ca_2+-binding protein whose binding to other proteins depends on the intracellular Ca_2+ concentration; component of signal-transduction pathways.

Calvin cycle (Calvin–Benson cycle, carbon reduction cycle, C_3 cycle):

- Pathway of photosynthetic CO_2 assimilation beginning with carboxylation of RuBP by Rubisco.
- The biochemical pathway for the reduction of CO_2 to carbohydrate. The cycle involves three phases; the carboxylation of ribulose-1,5-bisphosphate with atmospheric CO_2, catalyzed by rubisco, the reduction of the formed 3-phosphoglycerate to trioses phosphate by 3-phosphoglycerate kinase and NADP-glyceraldehyde -3-phosphate dehydrogenase, and the regeneration of ribulose-1,5-bisphospate through the concerted action of ten enzymatic reactions.
- First stable compound formed after the entry of CO_2 in plant system is triose sugar viz. Phosphoglyceric acid (PGA) (Calvin, 1961).

CAM (Crassulacean Acid Metablism) plants :

- Stomatal opening in night due to accumulation of malic acid, acts as strong solute and closing during day as a result of disappearance of malic acid from stomatal guard cells. Examples - Plants of arid region like Agave, Kalanchoe, sedum.
- Plants that fix CO_2 during the night into a four-carbon compound (malate) that, after storage in the vacuole, is transported out of the vacuole and decarboxylated during the day. The CO_2 released is assimilated by the Calvin cycle in the chloroplast stroma.

Cambium (Vascular cambium) :

- Layer of meristematic cells between the xylem and phloem that produces cells of these tissues and results in the lateral (secondary) growth of the stem or root.
- Thin layer of cells associated with secondary growth.

Canavanine : Toxic, nonprotein amino acid that is a close analog of the protein amino acid arginine.

Canopy conductance/resistance : Conductance/ Resistance for transport of CO_2, water vapour or heat between the plant canopy and the atmosphere measured across the boundary layer of the canopy.

Capillarity :

- The movement of water along a capillary tube due to cohesion, adhesion and surface tension.
- The movement of water for small distances up a glass capillary tube or within the cell wall, due to water's cohesion, adhesion, and surface tension.

Capillary tube : A glass tube with narrow bore.

Capillary water :

- Water fills the spaces between noncolloidal smaller soil particles and forms films around them. Available for plant growth.
- A form of water in the seed occurring at 10–20% seed moisture content.

Carbamylation : Reaction between CO_2 and an amino group; in many species, Rubisco is activated by carbamylation, catalyzed by Rubisco activase.

Carbohydrates : Polyhydroxy aldehydes or ketones or substances that yield such compounds on hydrolysis. The ratio of carbon to hydrogen to oxygen atoms is 1 : 2 : 1. Its emprical formula is $(CH_2O)n$.

Carbon & energy source : Usualy glucose, sucrose and fructose.

Carbon dioxide compensation point : The CO_2 concentration (partial pressure of CO_2 in the intercellular air space) at which CO_2 fixed by photosynthesis and CO_2 produced by respiration equal each other.

Carbonic anhydrase : Enzyme catalyzing the interconversion of HCO_3 - and CO_2.

Carbowax : A chemical used for seed priming; commercially known as PEG 6000.

Carboxylate : Organic acid minus its protons.

Carboxylation :

- Binding of a CO_2 molecule to a CO_2-acceptor molecule.
- The reaction in which a carboxylase catalyzes the formation of a carbon bond between CO_2 and the carbon atom of an organic molecule.

Carboxylation efficiency : Initial slope of the CO_2 response curve of photosynthesis.

Cardenolides : Steroidal glycosides that taste bitter and are extremely toxic to higher animals through their action on Na^+K^+-activated ATPases. Extracted from foxglove (Digitalis) for treatment of human heart disorders.

Cardinal temperature : Range of temperature characteristic for a particular species during which the germination is maximum.

Carotenoids :

- Accessory photosynthetic pigment; carotenoids of the xanthophyll cycle play a role in dissipation of excess energy.
- Lipid compounds, yellow to purple colour found in animals and plants both, cyanobacteria, photosynthetic bacteria, fungi etc. Associated with transfer of light energy to chlo. a and protection of chlorophyll against photooxiation by forming epoxy ring.

- Linear polyenes arranged as a planar zigzag chain, with the repeating conjugated double bond system- $CH{=}CH{-}CCH_3{=}CH{-}$. These orange pigments serve both as antenna pigments and photoprotective agents.

Carpels : The structure that contains the ovules of flowering plants, consisting ovary, style, and stigma. The carpel develops into the fruit and the ovule develops into the seed.

Carrier :

- Protein involved in ion transport across a Membrane.
- Membrane - transport proteins that bind to a solute, undergo conformational change, and release the solute on the other side of the membrane.
- Proteins present in the membrane. During transport, the substances being transported is initially bound to a specific site on the carrier protein which was released free on the inner side of membrane enzymatically.

Carry-over seed : The extra quantity of seed of foundation or certified seed that is carried over from previous crop season and used in case of failure current season seed crop.

Carrier-mediated transport : Active or passive transport of a solute accomplished by a carrier.

Caruncle (¼strophiole) :

- An outgrowth of a seed coat, near the hilum; preformed weak site in the seed coat.
- A fragile appendage of the outer integument noticed in some seeds near the hilum. e.g. Castor, Rubber, Tapioca.

Cascade : A succession of interactions so that each interaction derives from or acts upon the product of the preceding. A series of interactions that involves an amplification of the initial action.

Casparian band/strip :

- Waxy suberin impregnation on the radial and transverse wall of endodermis and exodermis cells that renders the wall impermeable to water.

- Band of radial cell walls in the endodermis, impreguated with wax like hydrophobic substance subesin which acts as barrier to water or solute movement.
- A band in the cell walls of the endodermis that is impregnated with the waxlike, hydrophobic substance suberin. Prevents water and solutes from entering the xylem by moving between the endodermal cells.
- Thickened endodermal cell layer due to deposition of suberin. Structure prevents the apoplastic movement of water or solutes across the root.

Catabolic process : Pertaining to or characterized by the release of energy.

Cation : Positively charged ion.

Cation exchange : The replacement of mineral cations adsorbed to the surface of soil particles by other cations.

Cation exchange capacity : The degree to which a soil can adsorb and exchange ions.

Cauliflower mosaic virus (CaMV) : Virus used to deliver foreign DNA in to plant cells.

Cavitation :

- Breakage of a water column in a xylem conduit due to air seeding.
- The collapse of tension in a column of water resulting from the indefinite expansion of a tiny gas bubble.

CCCP (Carbonyl cyanide mchlorophenylhydrazone) : An ionophore that makes the plasma membrane highly permeable to protons, thus abolishing the proton gradient.

Cdc2 protein : The major cyclin-dependent protein kinase, cell division cycle 2. Biosynthesis stimulated by auxin.

Cell number : A convenient means to measure the growth of unicellular organisms. In multicellular plants, increasing numbers of cells are usually associated with cell expansion, particularly in meristems.

Cell plate : Wall like structure that separates newly divided cells. Formed by the phragmoplast and later becomes the cell wall.

Cell suspension culture :

- Culture of isolated cells or very small cell aggregates remaining dispersed in liquid medium.
- Shaking and growing the plant cells or cell groups in liquid medium.

Cell wall :

- Layer surrounds the cell in plants.
- Structural matrix surrounding plant cells; part of the apoplast.

Cell wall matrix : Plant cell wall material consisting of hemicelluloses and pectins plus a small amount of structural protein.

Cellobiose : A1>4-linked â-D-glucose disaccharide that makes up cellulose.

Cellular respiration : Respiration takes place in each cell in multicellular organisms. Each cell obtains its own energy.

Cellulase : Enzyme used for wall degradation to release protoplast.

Cellulose :

- A polymer of carbohydrate, constituent of cell wall.
- Glucose molecules joined by (1+ 4), glycosidic bonds.
- Linear chains of 1? 4 - linked â - D-glucose. The repeating unit is cellobiose.
- Linear unbranched homopolysaccharides of 10000 or more 0glucose units connected by 1-4 linkage (glycosydic bond). The 1-4 linkage are found in ß-configuration.
- Structural polymer of glucose; major component of plant cell walls giving tensile strength.

Cellulose microfibrill :

- Thin, ribbon-like structure of indeterminate length and variable width composed of 1>4-linked â-D-glucan chains tightly packed in crystalline arrays alternating with less organized amorphous regions. Provides structural integrity to the cell walls of plants and determines the directionality of cell expansion.
- Stiff structures that contribute to the strength and structural bias of cell wall.

Cellulose synthase : Enzyme that catalyzes the synthesis of individual 1>4-linked β-D-glucans that make up the cellulose microfibril.

Cell-wall elasticity : Reversible change in cell-wall Dimensions.

Cell-wall extensibility : Irreversible extension of cell walls, due to structural changes.

Central zone :

- A central cluster of relatively large, highly vacuolate, slow-dividing cells in shoot apical meristems, comparable to the quiescent center of root meristems.
- Center of an active meristem which contains a cluster of relatively large, highly vacuolate cells.

Centrifugal : Progress of anthesis from the centre towards either side of an inflorescence.

Cereal grains : Seeds of grasses consisting of the diploid embryo, the triploid endosperm and the fused seed coat-fruit wall.

Certified seed : Progeny of foundation seed or certified seed itself. It is the last step in generation system of seed multiplication. The production is supervised by certification agency. Tag colour-azure blue.

CF_0F_1 ATPase : A multi-protein complex associated with the thylakoid membrane that couples the passage of protons across the membrane to the synthesis of ATP from ADP and phosphate. Similar to F_0F_1ATP synthase in oxidative phosphorylation but much less sensitive to oligomycin.

Chaffy Seed : The seed attached with light appendages. e.g. Grasses.

Chalaza : Lower end of embryo sac.

Channels :

- Transmembrane proteins that function as selective pores for passive transport of ions or water across the membrane.
- Transmembrane proteins that function as selective pores, through which molecules or ions can diffuse across membrane channels only permit passive absorption and limited mainly to ions or water.

Chaperones : Group of stress proteins that are encoded by a multigene family in the nucleus; chaperones bind to and stabilize an otherwise unstable conformation and, thus, mediate the correct assembly of other proteins.

Chelate :

- Compound that combines reversibly, usually with high affinity, with a metal ion (e.g., iron, copper, or calcium).
- Substances such as EGTA that form a complex with divalent cations eliminating their biological activity.

Chelators :

- Carbon compounds that can form a noncovalent complex with certain cations facilitating their uptake (e.g., malic acid, citric acid).
- Cation-binding organic molecule, such as citric acid, malic acid, and phytosiderophores.
- Compounds which form soluble complexes with cations such as iron and calcium in which the cation is held by ionic forces thus are physically more available to a plant. Examples - Ethylenediaminetetraacetic acid (EDTA) or Diethylene-triaminepentaacetic acid (DTPA or pentetic acid).

Chemical potential :

- Amount of free energy, kinetic energy or translational energy available in a system.
- The free energy associated with a substance that is available to perform work.

Chemical potential gradient : A change in the free energy per mole of a substance, measured over a given distance. A substance moving spontaneously does so down its chemical potential gradient.

Chemical potential of solute : Sum of concentration, electric and hydrostatic potentials.

Chemical Potential of given solute under standard conditions	=	Chemical potential of solute	+	Concentration (activity) component	+	Electric potential component	+	Hydrostatic pressure component

Chemical scarification : Scarification of seeds done using strong acid, alkali or oxidants.

Chemically combined water : Water bounds to soil minerals by strong chemical bonds and not available for plants.

Chemiosmotic mechanism : The mechanism whereby the electrochemical gradient of protons established across the thylakoid membrane by the electron transport process between photosystems II and I is used to drive energy requiring cellular processes such as ATP synthesis. It also operates in mitochondrial respiration and at the cell plasma membrane.

Chemiosmotic model : Theory accounting for the synthesis of ATP driven by a proton-motive force.

Chemosynthetic Bacteria : Manufacture of food material by certain arobic colourless bacteria by using chemical energy released during biological oxidation of certain inorganic substances. Examples - Nitrosomonos, Nitrobacter, Spirophyllum, Bacillum pantotrophus etc.

Chilling injury/ tolerance :

- Injury caused by exposure of plants or tissues to low temperatures (> 0°C); tolerance of such temperatures.
- Changes that occur when plants growing at 25 to 35°C are cooled to 10 to 15°C. Includes slowed growth, leaf discolouration and / or lesions. Contrast with freezing injury.
- Damage caused to seed on exposure to low temperature (10-15°C).
- When plants growing at relatively warm temperatures (25 to 35°c) are cooled to 10 to 15°c. Injury is associated with slower growth, discoluoration or lesions appear on leaves and the foliage looks soggy, as if soaked in water for a long time. If roots are chilled the plants may wilt.

Chilling sensitivity : Seeds with high moisture are damaged when exposed to low temperature due to accumulation/formation or ice crystals as in case of recalcitrant seeds.

Chilling-sensitive plants : Plants that experience a sharp reduction in growth rate at temperatures between 0 and 12°C.

Chimera : Mixture of cells with a different genetic make up. When a mutation occurs in a stem cell in the shoot apical meristem, all the cells in the plant derived from the stem cell will carry mutation.

Chitin : Polymer of N-acetylglucosamine; component of the exoskeleton of arthropods and the cell wall of fungi, but not of plants.

Chitinase : Chitin-hydrolyzing enzyme that breaks down fungal cell walls.

Chlorenchyma : Tissue containing chloroplasts.

Chlorophylls :

- A group of light absorbing green pigments active in photosynthesis.
- Green pigment in the photosynthetic membrane (thylakoid) involved in light capture as the first step in photosynthesis.
- Pigments associated with light absorption during photosyntghesis. Nine types have been identified viz., Chloro a, b, c, d, e, bacterochlorophyll a & b and chlorobium chloro. 650 and 660 mμ.

  Empirical formula Chlo. a – $C_{55} H_{72} 0_5 N_4 Mg$
  Chlo. b – $C_{55} H_{70} O_6 N_4 Mg$

Chlorophyll deficiency complementation : Isolation of somatic hybrids after fusion of two distinct homozygous recessive albino mutants using chlorophyll deficient light sensitive varieties subthal and virescent.

Chlorophyllase : An enzyme that removes the phytol from chlorophyll as part of the chlorophyll breakdown process.

Chlorophyte : Unicellular photosynthetic eukaryote whose chloroplasts contain chlorophyll a and b (green algae).

Chloroplasts :

- Membranes contain chlorophyll and its associated proteins,, sites of photosynthesis.
- The organelle that is the site of photosynthesis in eukaryotic photosynthetic organisms.

- Spherical structures contain carbohydrates, proteins, lipids, chlorophylls and carotenoids. Light and dark reactions occur in grana and stroma, respectively.

Chlorosis :

- The yellowing of older, lower plant leaves characteristic of prolonged nitrogen deficiency.
- Yellowing of leaves due to loss of chlorophyll.

Chlorox soak test : A method for determining soybean or bean seed processing damage by soaking in cholorox (liquid bleach) and counting the swollen seeds.

Cholodny-Went model : Early mechanism proposed for tropisms involving stimulation of the bending of the plant axis by lateral transport of auxin in response to a stimulus, such as light, gravity, or touch. The original model has been supported and expanded by recent experimental evidence.

Cholorotic seedlings : Seedlings without chlorophyll.

Chromatin : DNA + protein complex.

Chromatophore : A light absorbing pigment molecule, one of the components of phytochrome.

Chromophore :

- A light absorbing pigment molecule that is usually bound to a protein (an apoprotein).
- Light-absorbing constituent of a macromolecule (photoreceptor) that is responsible for light absorption.

Chromoplasts :

- Pigmented plastids contain carotenoids, phycoerythrin and fucoxanthin.
- Plastids contain high concentrations of carotenoids, impart yellow, orange or red colours to many fruit and flowers.

Chromosomes : Net work of fibre resemble structures containing genetic informations.

Chromosomal aberrations :

- Deviations or alternations in the chromosomes.

- Change in chromosome structure.
- Deviations from normal chromosome structure.

Chronic photoinhibition : Photoinhibition of photosynthetic activity in which both quantum efficiency and the maximum rate of photosynthesis are decreased. Occurs under excess light.

Cinnamic acid : A phenyl-propanoid derived from the amino acid phenylalanine that is a key intermediate in the biosynthesis of many phenolic compounds.

Circadian rhythms :

- A biological activity that shows a cycle of high activity and low activity independent of external stimuli, with a regular periodicity of about 24 hours (L. circa diem : about a day).
- Occurrence of various metabolic processes in plants through high activity and low activity phases with a regular periodictiy of about 24 hours.

Cis and trans-acting : DNA binding protein.

Cis-acting regulatory elements : Upstream sequences, that regualate the amount and pattern of transcription.

Citric acid cycle (kreb's cycle, tricarboxylic acid cycle) : A cycle of reactions localized in the mitochondrial matrix that catalize the oxidation of pyruvate to CO_2, ATP and NADH are generated in the oxidation process.

CKI1 gene : Gene whose overexpression confers cytokinin-independent growth on Arabidopsis cells in culture. Encodes a protein similar to bacterial histidine kinases functioning in signal transduction.

Climacteric :

- Marked rise in respiration at the onset of ripening that occurs in all fruits that ripen in response to ethylene, and in the senescence process of detached leaves and flowers.
- Rise in respiratory rate before the ripening phase in response to ethylene.

Climax species : Species that are confined to later stages of succession in a plant community; as opposite to pioneer.

Clock hypothesis :

- Currently accepted hypothesis of how plants measure night length. Proposes that photoperiodic timekeeping depends on the endogenous oscillator of circadian rhythms.
- Photoperiodic timekeeping depends on an endogenous circadian oscillator of the type involved in the daily rhyhms in relation to phytochrome (Bunning 1960).

Clonal growth : Asexual production of physiologically complete plants; a form of vegetative reproduction.

Clone : Word derived from Greek (Clon=twig, spray or slip, like those broken off as propagules for multiplication), first used by webber to apply to cultivated plants that were propagated vegetatively.

Cloning : Multiplication.

Cluster roots : Bottle-brush-like or Christmas-treelike structures in roots with a dense packing of root hairs, releasing carboxylates into the rhizosphere, thus solubilizing poorly available nutrients (e.g., phosphate) in the soil.

CO_2-compensation point : CO_2 concentration at which the rate of CO_2 assimilation by photosynthesis is balanced by the rate of CO_2 production by respiration.

Coating : Covering seed with inert material such that the original shape of seed is retained.

Coconut milk : The liquid endosperm of coconut seeds that contains cytokinins and other nutritional factors. Stimulates the growth of normal stem tissues when added to liquid culture media.

Coefficient of velocity of germination : First suggested by kotowski (1926). It is the reciprocal of the average time to germination. The maximum CVG value is 100 if all seeds germinate on day one.

Coenzyme : A nonproteinaceous organic substance that combines with a specific protein, the apoenzyme.

Coenzyme A : A coenzyme with an -SH group that serves as an acyl group carrier for many enzymatic reactions.

Coevolution : Evolution of two (or more) species of which at least one depends on the other as a result of selection by mutual interactions.

Cofactor : Inorganic ion or coenzyme required for an enzyme's activity.

Cohesion – tension theory of sap ascent :

- Requires the cohesive properties of water to sustain large tensions in the Xylem water columns
- Upward movement of water is due to negative pressure, pull or suction created in leaves due to transpiration and cohesive force of water (Dixon 1910).
- A model for sap ascent in the xylem up the stem of the plant, stating that evaporation of water from the leaves at the top of the stem causes a tension (negative hydrostatic pressure) that pulls water up the long water columns in the xylem.
- Accounts for the ascent of sap in the xylem due to the cohesive forces between ascending water molecules under high tension and the adhesive forces between water and capillaries in the wall of xylem conduits.

Cohesive force of water : Attraction between water molecules responsible for forcing water upto a considerable height and surface tension which prevents the passage of air bubbles through minute pores and pits in cell walls.

Coincidence model : Model in the which the circadian oscillator controls the timing of light sensitive and light insensitive phases (Bunning 1960)

Colchicine :

- A drug that destroys microtubules and blocks cell division.
- Chemical used for chromosome doubling.

Cold stratification : A method to remove embryo dormancy for those seeds that have low GA levels while shedding; such seeds are placed in moist media at cold temperature (5-10^{0}C) which increases the endogenous production of GA; this increases in GA counteracts the effects of ABA and promotes germination. **Syn.** Prechil.

Cold test :

- A vigour test for assessing planting value of corn seeds using low temperature (10^0C) for seven days followed by germination at normal temperature.
- Special germination test; for corn, seeds are placed in a very moist sand/soil/peat moss mixture, then held at 10°c for 7 days before being placed in the regular germination temperature

Coleoptile :

- A modified ensheathing leaf that covers and protects the young primary leaves of a grass seedling as it grows through the soil. Unilateral light perception, especially blue light, by the tip results in asymmetric growth and bending due to unequal auxin distribution in the lighted and shaded sides.
- Sheath like leaf of grasses and other monocotyledons; it protects the delicate growing point as it emerges from the soil.
- Sheath which covers plumule in initial stage of germination.
- The sheath surrounding and protecting the shoot initial of the embryo and young seedling of monocots.

Coleorhiza :

- Sheath which covers radicle in initial germination phase.
- The sheath surrounding the root tip in *Poaceae (Gramineae)* equivalent to root cap in dicots.

Collenchyma : A specialized parenchyma with irregularly thickened, pectin-rich, primary cell walls that function in support in growing parts of a stem or leaf.

Colligative properties : Properties of solutions that depend on the number of dissolved particles and not on their chemical characteristics.

Colloidal system : Solution of two heterogenous phases normally solid dispersed phase and liquid dispersion medium having low osmotic pressure (Greek word Kello - glue, Eidos like). Particle size range .001 - .1μ.

Columella : Central cylinder in root.

Columella root cap : The central region of the root cap that contains the statocytes-cells containing large, dense amyloplasts that function in gravity perception during root gravitropism.

Columella stem cells : Root cap stem cells that divide to generate a sector of the root cap, the columella.

Combinatorial model :

- Model which accounts for intermediate forms during the transition from juvenile to adult in maize.
- Proposal that during the transition from juvenile to adult shoot in maize, a series of independently regulated, overlapping programs (juvenile, adult, and reproductive) modulate the expression of a common set of developmental processes.

Combined dormancy : Dormancy caused by two primary factors such as seed coat and embryo dormancy. **Syn.** Double dormancy.

Combuston and respiration : In combuston the oxidation of substrate is sudden, whereas in respiration it is step wise.

Companion cells :

- Cell type in the phloem, adjacent to sieve element, involved in phloem loading.
- In angiosperms, a metabolically active cell that is connected to its sieve element by large, branched plasmodesmata and takes over many of the metabolic activities of the sieve element. In source leaves, it functions in the transport of photosynthate into the sieve elements.

Compartmentation :

- Restriction of compounds or processes to specific cells, or parts of a cell, such as storage of secondary metabolites in vacuoles.
- Compartments in vivo because the site of the peak in a spectrum depends on the pH around the molecule.

Compatible interaction : Response of a susceptible host to a virulent pathogen; positive interaction between pollen and pistil allowing guidance of the sperm cells toward the ovule.

Compatible solutes (compatible osmolytes) :

- Solute that has no deleterious effect on metabolism at high concentrations Compensation point conditions (temperature, [CO_2], light) where net CO_2 exchange by a leaf or plant is zero (i.e., photosynthesis equals respiration).
- Organic compounds that are accumulated in the cytosol during osmotic adjustment. Compatible solutes do not inhibit cytosolic enzymes as do high concentrations of ions. Examples of compatible solutes include proline, sorbitol, mannitol, glycine betaine.
- Organic compounds that do' not interfere with the enzyme functions.

Competence : The capacity of a particular cell or group of cells to respond in the expected manner when given the appropriate developmental signal.

Competent bud : Bud which is able to flower when given the appropriate developmental singal.

Competition : Interaction among organisms (of the same or different species), which utilize common resources that are in short supply (resource competition), or which harm one another in the process of seeking a resource, even if the resource is not in short supply (interference competition).

Competitive ability : Probability of winning in competition with another species in a particular environment.

Complementation of resistance markers : Recognisition of dominant character such as traits conferring resistance to antibiotics, herbicides, amino acid analogues or other toxic compounds as potent selective markers.

Complex I : A protein complex in the mitochondrial electron transport chain that oxidizes NADH and reduces ubiquinone.

Complex II : A protein complex in the mitochondrial electron transport chain that oxidizes succinate and reduces ubiquinone.

Complex III : A protein complex in the mitochondrial electron transport chain that oxidizes reduced ubiquinone (ubiquinol) and reduces cytochrome c.

Complex IV : A protein complex in the mitochondrial electron transport chain that oxidizes and reduced cytochrome C and reduces O_2 and H_2O.

Compound (or mixed) fertilizers :

- Contain two or more mineral nutrients; numbers such as 10-14-10 refer to the effective percentages of nitrogen, phosphorus, and potassium.
- Fertilizers that contain two or more mineral elements.

Compound lipids : Contains in their molecules other groups eg. phospholipids which contain phosphoric acid.

Concentration of solution : It gives the strength of solution expressed on volume, weight or % basis.

Condensed tannins : Tannins that are polymers of flavonoid units. Require use of strong acid for hydrolysis.

Conditioning : Refers to the removal of awn, wings and other appendages and preparing the seed for further processing.

Conductance : Flux per unit driving force (e.g., concentration gradient); inverse of resistance.

Conductivity test : A vigour test that measures the electrical conductivity of seed steep water.

Constitutive :

- Constantly present or expressed, whether there is demand or not. Refers to the ongoing synthesis of a particular protein. Contrast with inducible.
- Produced in constant amount (as opposed to regulated) (e.g., genes can be expressed constitutively).

Constitutive defense : Background level of plant defense in the absence of induction by herbivores or pathogens.

Constitutive ethylene response mutants : Mutants that show the ethylene triple response in the absence of exogenous ethylene.

Construction cost : Carbon and nutrients required to produce new tissue, including the respiration associated with the biosynthetic pathways.

Contact exchange :

- Exchange of ions of the same charges between two surfaces, root surface and clay particles.
- Exchange of ions of the same charge held on the surface of soil colloids or root surface.

Containers (as per Seeds Act) : Refers to a box, casket, tin, wooden case, receptacle, plastic, sack, cloth bag, paper packet or other packages used for carrying or storing seeds.

Continuous cultures : Draining out of used material and adding fresh medium.

Continuous drying : Drying carried out with continuous flow of grain and air, contrast to bath drying.

Contractile roots : Mature roots that decrease in length, while increasing in diameter, thus pulling the plant deeper in the soil, as in geophytes.

Controlled atmosphere storage : A method of storage where the storage atmosphere is modified to prevent loss of seed viability by enriching with inert gases like Nitrogen or Carbon di-oxide.

Controlled Deterioration test (CD Test) : A modified accelerated ageing vigour test where the seed moisture is raised to specific level before subjecting to accelerated ageing; this is quick and more accurate.

Convective heat : Transfer direct transfer of heat (e.g., from leaf to air) and further transport by turbulent movement.

Convergent : Evolution process whereby, in organisms that are not closely related, similar traits evolve independently as a result of adaptation to similar environments or ecological niches.

Cool germination test :

- Seeds are planted in moist towel paper or sand and incubated at 10-15° C upto final count day.
- A vigour test used to assess the planting value of cotton seeds using cool temperature (18°C) for seven days followed by germination under ambient conditions.

Cork cambium :

- A layer of lateral meristem that develops within mature cells of the cortex and the secondary phloem. Produces the secondary protective layer, the periderm.
- Meristematic layer that develops within mature cells of the cortex and the secondary phloem.

Cork formation in Apple : Deficiency of Boron results in hardening of central pith in apple. Fruit becomes brittle and dry resulting in reduced fruit size.

Corpus : The internal cytohistological zones of the shoot apical meristem : central zone, peripheral zone, rib meristem.

Cortex : Few layers of parenchymatous cells internal to epiblema contain intercellular spaces.

Cortical cytoplasm : The outer region or layer of cytoplasm adjacent to the plasma membrane.

Cortical-endodermal stem cells : A ring of stem cells that surround the quiescent center and generate the cortical and endodermal layers in roots.

Cotransport : The simultaneous transport of two solutes by the same carrier. Usually one solute is moving down its chemical-potential gradient, while the other is moving against its chemical-potential gradient.

Cotyledons :

- Seed leaves in dicot embryo used as storage organ and serves as assimilatory organs in some plants.
- Seed leaves of the embryo usually thickened for storage of reserve food; they may serve as true foliage leaves.
- The one or more seed leaves contained in the seed of seed plants. In some seeds they are storage organs supporting early non-photosynthetic growth of the seedling. In other seeds, they absorb and transmit to the seedling resources stored in the endosperm.

Coumarins : A group of phenylpropanoid compounds including the phototoxic furanocoumarins, and other substances responsible for the odor of fresh hay.

Coupling : A process by which a chemical reaction releasing free energy is linked to a reaction requiring free energy.

Coupling factor : ATP synthetase in thylakoid membrane of chloroplasts and inner membrane of mitochondria.

C_pT_i : Gene for cowpea trypsin inhibitor.

Cracked cotyledons : A type of processing damage in beans leading to cracked cotyledons.

Cracked seeds : Seeds with cracked seed coat caused by processing damage.

Crassulacean acid metabolism (CAM) :

- A biochemical process for concentrating CO_2 at the carboxylation site of rubisco. Found in the family Crassulaceae (*Crassula, Kalanchoe, Sedum*) and numerous other families of angiosperms. In CAM, CO_2 uptake and fixation take place at night, and decarboxylation and reduction of the internally released CO_2 occur during the day.
- Photosynthetic pathway in which the first step of CO_2 assimilation is the carboxylation of phosphoenolpyruvate (PEP) by PEP carboxylase; the first product is oxaloacetic acid (OAA)– a four-carbon intermediate; in contrast to C_4 photosynthesis, in CAM photosynthesis, the CO_2 assimilation occurs predominantly during the

CRE1 gene : Arabidopsis gene that encodes a cytokinin receptor protein, similar to bacterial two-component histidine kinases.

Cristae (singular crista) :

- The invaginations of inner membrane in mitochondria.
- The infolding of inner membrane in mitochondria.
- Folds in the inner mitochondrial membrane that project into the mitochondrial matrix. The enzymes of the electron transport chain and of oxidative phosphorylation are localized in the cristae.

Cristanae : Flattened or tubular sacs composed of membranes.

Criteria of essentiality of elements : Arnon and Stout (1939) and Epstein (1972) gave the criteria to designate an element as an essential element.

Critical concentration (of a nutrient) :

- The minimum tissue content of a mineral nutrient that is correlated with maximal growth or yield.
- Minimum content of nutrient in tissue that is correlated with maximal growth or yield.

Critical day length : The minimum length of the day required for flowering of a long-day plant; the maximum length of day that will allow short-day plants to flower. However, studies have shown that it is the length of the night, not the length of the day that is important.

Critical LAI : LAI values required for maximum crop growth rate with 95% light interception.

Critical moisture limit : Refers to the minimum moisture limit of recalcitrant seeds to which they can be dried. Drying below this causes death. The moisture content at which the germination is 60%.

Critical night length : The night length must be exceeded for flowering of short-day plants, or for inhibition of flowering in long-day plants.

Critical oxygen pressure (COP) : Oxygen pressure at which the respiration rate is first slowed by O_2 deficiency.

Crop growth rate (CGR) : Daily increment in plant biomass per unit groud area per unit time.

Watson (1952) - $CGR = \frac{W_2 - W_1}{P(t_2 - t_1)}$ Weight $Area^{-1}$ $Time^{-1}$

Where P = ground area which a plant occupies or group of plants occupy.

Cross protection : Inoculation of susceptible strain of a crop with a mild strain of a virus to develop resistance against more virulent strains.

Cross-resistance : The phenomenon in which an organism that has acquired resistance to one pathogen or herbivore through direct exposure simultaneously has acquired resistance to other pathogens or herbivores to which it has not been exposed. Cross-resistance arises because the biological mechanism of resistance is the same and arises through identical genetic mutations.

Cross-talk : Communication between different signal transduction pathways.

Crown gall : A tumour-forming plants disease resulting from wound infection of the stem or trunk by the soil-dwelling bacterium *Agrobacterium tumefaciens.* A tumour resulting from the disease.

Crown roots : Adventitious roots originates from crown region.

CRY1 gene : The gene encoding cryptochrome, a flavoprotein implicated in many blue light responses that has homology with photolyase. Formerly HY4.

cry1 mutant : Arabidopsis mutant that lacks the blue light-stimulated inhibition of hypocotyl elongation.

Cryogenic storage : Refers to storage at very low temperatures. e.g. Liquid nitrogen (-196^{0}C).

Cryopreserved embryos : Embryos preserved at cryogenic temperatures or in liquid nitrogen (-196^{0}C).

Cryoprotectant : Chemicals that prevent the damage to seeds subjected to very low temperatures (-196^{0}C) by preventing ice crystal formation. e.g. Dimethyl sulfoxide (DMSO), Ethylene glycol, Sucrose, L-proline, Glycerol.

Cryptochrome :

- A group of unidentified pigments that absorb blue light and also long wave ultravoilet wavelengths (the UV - A region, about 320 to 400 nm). found in cryptogams (non flowering plants).
- Blue-light-absorbing photoreceptor, involved in photo-morphogenesis.
- Flavoprotein involved in the inhibition of stem elongation, possibly as a blue-light receptor or as a signaling intermediate.

Cryptochrome 1 (Cry 1) : Blue light photoreceptor mediating the inhibition of stem elongation.

Crystalline : Pertaining to solids having a highly ordered and repetitive geometric form.

CTR1 gene : Encodes for a negative regulator of ethylene responses.

CTR1 mutant : In *Arabidopsis,* a recessive mutation causing the constitutive expression of the ethylene triple responses (constitutive triple response 1 = triple response in the absence of ethylene).

CTR1 protein : Regulator of the ethylene triple response; resembles serine/threonine protein kinases involved in signal transduction.

Cultivar : Cultural or cultivated variety.

Culture facilities : Controlled temperature, illumination, photoperiod, humidity air circulation devices with incubators, growth chambers, walkin environmental rooms etc. for tissue culture work.

Culture medium : Laboratory substance on which microorganisms and excised embryos may be grown.

Curvature test : A bioassay for auxin using the curvature of the Avena coleoptile in response to asymmetrically applied auxin in an agar block.

Cutan : A lipid polymer made up of long-chain hydrocarbons that is a constituent of the cuticle.

Cuticle :

- A multilayered structure that coats the outer cell walls of the epidermis and restricts the passage of water and gases into and out of the plant. Includes cutin, cutan, and waxes.
- Very thin detachable skin covering a plant, especially the leaves.
- Waxy coating of external plant surfaces.

Cuticular conductance/resistance : Conductance/ resistance for transport of CO_2 or water vapour movement through the cuticle.

Cuticular transpiration : Loss of water by transpiration from outer walls of epidermal cells through cuticle.

Cutin :

- Macromolecule, a polymer consisting of many long chain fatty acids attached to each other by ester linkages creating a rigid three dimensional network. It is peripheral constituent of the cuticle, a multilayered secreted structure that coats the outer cell walls of the epidermis on the aerial parts of all herbaceous plants.

- A rigid three-dimensional polymer of hydroxyl-bearing fatty acids that are attached to each other by ester linkages. The principal constituent of the cuticle.
- A type of compound, waxy in nature that coats the seed coat, leaves and protects them against insects, fungi, water etc.
- Found in outer walls of epidermal cells of plants forms a thick layer called cutilce.
- Waxy substances that are part of the cuticle; polymer consisting of many long-chain hydroxy fatty acids that are attached to each other by ester linkages, forming a rigid threedimensional network.

Cyanide : An inhibitor of complex IV and other heme-containing enzymes.

Cyanogenic : Releasing cyanide.

Cyanogenic glycosides :

- Nitrogenous protective compounds readily broken down to release the well known poisonous gas hydrogen cyanide (HCN).
- Nonalkaloid, nitrogenous protective compounds that break down to give off the poisonous gas hydrogen cyanide when the plant is crushed.

Cybrids : Fusion of two protoplasts which combines both parents cytoplasms with only one parents nuclear genome.

Cyclic electron flow : In photosystem, flow of electrons from the electron acceptors through the cytochrome b_6f complex and back to P700, coupled to proton pumping into the lumen. This electron flow energizes ATP synthesis but does not oxidize water or reduce NADP+.

Cyclic Photophosphorylation : Photosystem I (PSI) gets back electrons in the cyclic fashion.

Cyclin dependent protein kinases (CDKs) :

- Enzymes involved in regulation of cell cycle.
- Protein kinases that regulate the transitions from G1 to S, and from G2 to mitosis, during the cell cycle.

Cyclins : Regulatory proteins associated with CDKs that play a crucial role in regulating the cell cycle.

Cycloheximide : Inhibits eukaryotic protein synthesis on 80S ribosomes, but does not block protein synthesis in prokaryotes, mitochondria, or chloroplasts.

Cyclosis or protoplast streaming : Test used as a measure of protoplast viability.

Cytochalasin B : A drug that destroys actin filaments.

Cytochrome : Colored, heme-containing protein that transfers electrons in the respiratory and photosynthetic electron transport chain.

Cytochrome b_6 f complex : A large multi-subunit protein containing two b-type hemes, one c-type heme (cytochrome f), and a Rieske iron-sulphur protein. A nonmobile protein distributed equally between the grana and the stroma regions of the membranes.

Cytochrome C : A peripheral, mobile component of the mitochondrial electron transport chain that oxidizes complex III and reduces complex IV.

Cytochrome oxidase : Mitochondrial enzyme catalyzing the final step in the transfer of electrons from organic molecules to O_2.

Cytochrome P450 : Element in the synthesis of anthocyanins and in the detoxification of xenobiotics.

Cytochrome pathway : Phosphorylating electrontransport pathway in the inner membrane of plant mitochondria, transporting electrons from NAD(P)H or FADH2 to O_2, with cytochrome oxidase being the terminal oxidase.

CytoHistological zonation : Organisational pattern of active apical meristem.

Cytohistological zones : Regions of the shoot apical meristem showing differences in morphological appearance and rates of mitosis.

Cytokinesis :

- In plant cells, following nuclear division, the separation of daughter nuclei by the formation of new cell wall.
- Process whereby a cell is partitioned into two progeny cells.

Cytokinin oxidase :

- Enzyme that inactivates cytokinins by removing the isoprene moiety from adenine or its derivatives.
- Enzyme which cleaves the side chain from zeatin (both cis and trans), zeatin riboside, ip and their N-glycosides, but not their glucoside derivatives. However, dihydrozeatin and its conjugentes are resistant to cleavage.

Cytokinin synthase : Plant enzyme that transfers the isopentenyl group from iso[entenyl diphosphate to AMP to form isopentenyl adenine ribotide, the first unique intermediate in the synthesis of cytokinin. An isopentenyl transferase.

Cytokinin(s) :

- A class of phytohormones, involved, e.g., in the delay of leaf senescence, cell division, cell extension, release of dormancy of buds, and chloroplast differentiation.
- Compounds with many developmental effects on plants, including leaf senescence, nutrient mobilization, apical dominance, the formation and activity of shoot apical meristems, floral development, the breaking of bud dormancy, and seed germination. Cytokinins mediate many light regulated processes, including chloroplast development and metabolism, and expansion of leaves and cotyledons. May exist as free and bound forms. Operationally defined as compounds with biological activities similar to those of trans-zeatin.
- Substances enhance cell division in combination with auxins and delays leaf senescence. Kinetin, zeatin and benzyladenine are main cytokinins.
- Derivatives of adenine in which the H on the N_6 position of the ring is subsituted by either benzyl group forming benzylademine, furfuryl group forming kinetin or isopentenyl group forming zeatin.

Cytological change : Change in the intrinsic character or function of a cell.

Cytoplasm :

- Contents of a cell that are contained within its plasma membrane, but outside the vacuole and the nucleus.

- The cellular matter enclosed by the plasma membrane exclusive of the nucleus, that contains the cytosol, ribosomes, and the Cytoskeleton which, in eukaryotes, surrounds intracellular and membrane limited organelles (chloroplasts, mitochondria, endoplasmic reticulum, etc.).

Cytoplasmic male sterility (CMS) :

- A plant trait caused by mutations in mtDNA in which viable pollen is not formed.
- Sterility induced by obtaining cell organelles from different species. Examples - Brassica raphanus cybrids that contain the nucleus of B. napus, chloroplasts of atrizine resistant B.campestris and mitochondria that confer male sterility from Raphanus sativus.
- Plants that do not form viable pollen, a physiological feature directly linked to the plant mitochondrial genome.

Cytoskeleton : Organised cytosol into a three dimensional network of filamentous proteins.

Cytosol :

- Cellular matrix in which cytoplasmic organelles are suspended.
- The colloidal-aqueous phase of the cytoplasm containing dissolved solutes but excluding supramolecular structures, such as ribosomes and components of the cytoskeleton.

❑❑❑

D

Damaged seedling :

- Seedling with any of the essential structures missing or so badly damaged that balanced development is affected.
- Seedlings with any of essential structures missing.

Damping-off : A condition in which young seedling are attacked and killed by soil borne fungi following germination.

Damp seed heating : Increase in temperature of stored high moisture seeds due to metabolic activity and associated fungal growth.

Dark reaction : Carbon fixation during photosynthesis; does not directly require light but uses the products of the light reaction.

Dark respiration : Processes in the cytosol, plastids, and mitochondria that break down carboncontaining compounds and generate ATP; it produces CO_2 and consumes O_2 when aerobic; when referring to gas exchange, all decarboxylation and O_2-consuming processes are included, apart form photorespiration.

Day neutral plants :

- Flowers after a period of vegetative growth regardless of photoperiod. Examples - Tomoto, Mirabilis, Peas, Balsam etc.
- A plant whose flowering is not regulated by day length.

DCMU (Dichlorophenyldimethylurea diuron) : An herbicide that blocks photosynthetic electron flow by displacing Q_B from the photosystem II reaction center complex. It also causes a partial inhibition of light stimulated stomatal opening.

De- differentiation : Process by which cells lose their differentiated characteristics, reinitiate cell division, and may, under appropriate conditions, regenerate whole plants.

de novo synthesis : Synthesis and /or assembly from simple molecular species.

Dead seeds : Refers to those seeds that have not germinated within the prescribed period of germination and can be identified by its decayed nature upon pressing between index finger and thumb.

Debearder : A pre-cleaning machine that removes awns in case of oats / barley and appendages of vegetable seeds (carrot); the fluffs of grass seeds in order to increase the processing efficiency.

Decayed seedling :

- A classification of abnormal seedling in which the seedling with any of their essential structures so diseased or decayed as a result of primary infection that normal development is prevented. This may be due to external or internal seed borne disease.
- An abnormal seedling in which the seedling with any of their essential structure diseased or decayed due to primary infection.

Deciduous : Having leaves that fall off or are shed seasonally in response to specific environmental cues, such as that occurs during or preceding unfavorable seasons.

Deciduous trees : Trees that shed their leaves seasonally.

Decomposition : Breakdown of organic matter through fragmentation, microbial and chemical alteration, and leaching.

Decorticated seed : Seed with its outer covering removed.

Decussate phyllotaxy : Phyllotactic pattern in which the two opposite leaves at one node are at right angles to the two leaves at the next node.

De-embryonate : To remove the embryo from a cereal seed, leaving only the starchy endosperm, aleurone layer, and fused seed coat fruit wall. De- embryonated half-seeds are dependent on exogenous GA to produce á-amylase.

Deep dormancy : Dormancy that requires extensive after ripening to overcome.

De-etiolation : Rapid developmental changes associated with loss of the etiolated form due to the action of light.

Defective seedling : A seedling that is damaged, deformed or decayed to an extent that normal development is prevented.

Defense compound : Secondary metabolite conferring some degree of protection from pathogens or herbivores.

Deficiency : while conducting a test in seed testing lab for purity, germination etc, an apparent deficiency exists if an estimate is less than a first estimate for a desired attribute or more for an undesirable attribute.

Deficiency zone :

- An increase in nutrient availability is directly related to an increase in growth or yield.
- Concentraations of a mineral nutrient in plant tissue below the critical concentration that reduces plant growth.

Deformed or unbalanced seedlings : Seedling with weak development or physiological disturbance in any essential structure.

Degree - day summation : A technique for judging that maturity of pine cone seeds based on the total temperature recorded during the maturation of cones. Higher sums lead to increased recovery of mature cones.

Dehumidification :

- Process by which the water is removed from a substance; moisture vapour is removed from air.

- The process of removal of moisture from seed using dry air of low relative humidity.

Dehumidified drying : A closed (air tight) system, where dehumidified air is circulated through the seed mass until seed moisture is reduced to desired level.

Dehumidifier : Equipment used to remove humidity (moisture) from air.

Dehydration response element (DRE) :

- A nine-nucleotide regulatory sequence element found in the promoters of genes regulated by abscisic acid (ABA), a plant hormone.

- Alternative set of proteins regulating transcription.

Dehydrins : Immunologically distinct family of proteins (Lea D11 family) that typically accumulate in plants during the late stages of embryogenesis or in response to any environmental influence that has a dehydrating effect.

Dehydrogenase : A group of respiratory enzymes found in seed, e.g. Succinic acid dehydrogenase, Malic acid dehydrogenase etc., that are closely related to life of a seed.

Dehydrogenation : Removal of hydrogen molecules from a compound.

Delayed greening : Pattern of leaf development typical of shade-tolerant rain-forest species; leaves are initially white, red, blue, or lightgreen during the stage of leaf expansion, reflecting their low concentration of chlorophyll and associated photosynthetic proteins.

Demand function : Dependence of net CO_2 assimilation rate on the intercellular or chloroplast CO_2 concentration, irrespective of the supply of CO_2 at ambient atmospheric CO_2 concentration.

Demand requirement : The term is used in the context of the control of the rate of a process (e.g., nutrient uptake, CO_2 assimilation) by the amount needed.

Denitrification :

- Conversion of soil nitrates into nitrogen and volatile nitrogen oxides by soil bacteria.

- Microbial conversion of nitrate to gaseous nitrogen (N_2 and N_2O); nitrate is used as an electron acceptor.
- Process by which soil anaerobic microbes convert nitrate (NO_3^-) or nirate (NO_2^-) to the gases, nitrous oxide (N_2O) and molecular nitrogen (N_2), which are then lost to the atmosphere.

Depelleted seeds : Pelleted seeds with coating removed for purity analysis, done by shaking the pelleted seeds on a wire mesh immersed in water to loosen off the pelleted material, the seeds are then cleaned and dried over a filter paper.

Deplasmolysis : Regaining of original state by plant cell or tissue as a result of water absorption.

Depolarization : Refers to a decrease in the usually negative membrane potential difference across the plasma membrane of plant cells. May be caused by the activation of anion channels and loss of anions, such as chloride from the cell interior, which is negative with respect to the outside.

Derived lipids : Hydrolytic products of naturally occurring lipids and include fatty acids, alcohals and sterols.

Dermal tissue system : Epidermis of tissue.

Desication tolerance - Ability of the cells to withhold water under stress

Desiccant : A chemical that is used to remove moisture from a seed sample or from air due to its hygroscopicity. **e.g.** Calcium chloride, Activated alumina, Silicagel, Phosphorous pentoxide.

Desiccation intolerant seeds : Seeds that loose viability on loss of moisture beyond a critical limit (usually 20-30%).

Desiccation postponement :

- Plant's ability to maintain tissue hydration when environmental water is limited.
- The ability to maintain tissue hydration.

Desiccation sensitive seeds : Syn. Recalcitrant seeds.

Desiccation tolerance (drought avoidance) :

- Plant's ability to function while dehydrated.

- The ability to function while dehydrated.
- Tolerance of extreme water stress, with recovery of normal rates of metabolism shortly following rehydration.

Desiccation tolerant seeds : Seeds that can withstand loss of moisture without affecting its viability.

Desiccator : Usually a glass jar fitted with an airtight cover and containing at the bottom some desiccating agent, such as calcium chloride.

Desmotubule : Narrow tubule of endoplasmic reticulum which a plasmodesma contains.

Desorption :

- Loss of moisture from a relatively moist hygroscopic material like seeds, to a relatively dry atmosphere till equilibrium is reached.
- Removal of a substance from an absorbed state. The reverse of adsorption.

Deterioration of varieties : A process either genetic or physiological that leads to reduction in varietal performance overtime.

Determinate growth :

- Apical meristem soon becomes differentiated in to an inflorescence and growth in length becomes stopped. Examples - most of the cereals.
- Inability to grow beyond the mature, reproductive stage due to the loss of meristematic activity. For example, plants that from flowers after the development of a preset numbers of leaves. Flowering times are generally short and all meristems of the plant are used up in the production of flowers.

Determination : Developmental state in which the plant follows a developmental program even after it is removed from its normal physical and environmental context.

Determined bud - A bud is said to be determined if it progresses to the next developmental stage (flowering) even after being removed from its normal context.

Development : Refers to the qualitative changes in plant parts. Example - Formation of flowers, fruits etc.

Devernalization : Erase of vernalization treatment by immediate exposure to higher temperature i.e. 30° C or above or with another exposure to low temperature.

Diaheliotropism :

- Leaf movements that maximize light interception by solar tracking and minimize overexposure to light.
- Solar tracking in which the leaf or flower remains perpendicular to incident radiation.

Diaphnoscope : An equipment for purity analysis, it has a hole (10 cm. dia.) at the center covered by etched glass. An electric bulb (40w) is fitted below the glass. Seeds are placed over the glass and observed for insect activity, purity components and admixture or red rice with white rice. Syn. Illuminated purity work board.

Dicarboxylic acid : Four or five carbon acids with two carboxyl groups. Examples - succinic acid, fumaric acid, malic acid etc.

Dicot : One of the two classes of flowering plants characterized by two seed leaves (cotyledons) in the embryo. Contrast monocot.

Dielatric constant : Measure of the capacity to neutralize attraction between electrical charges.

Dielectric properties : Properties that make a material a nonconductor of electricity.

Different plant tissue culture media : Nitsh and Nitsch, 1956; White (1963); Gautheret, 1942; Gamborgs (B_5), 1968; Chu (N_6), 1978.

Differentiation :

- Cellular specialization.
- Phenomenon of the reversion of immature cells to meristematic state leading to formation of callus.
- Process by which a cell acquires metabolic, structural, and functional properties that are distinct from those of its progenitor cell. In plants, differentiation is frequently reversible, when excised differentiated cells are placed in tissue culture.

- Process of specialization of cells to perform different functions. Examples - formation of xylem tracheids, sieve elements of phloem, guard cells of stomata from simple parenchymatus cells.
- The development of cells or tissues with a specific function or regeneration of organs like structures or (pro) embryos.

Diffuse growth : A type of cell growth in plants in which expansion occurs more or less uniformly over the entire surface. Contrast with tip growth.

Diffuse porous wood : In which wide and narrow xylem vessels are randomly distributed throughout each annual growth ring.

Diffusion :

- Random movement of ion, molecule or atom from area of higher concentration to lower concentration or from area of high kinetic energy to low kinetic energy influenced by their own kinetic energies.
- Net movement of a substance along a concentration gradient due to random kinetic activity of molecules.
- The movement of substances due to random thermal agitation from regions of high free energy (high concentration) to regions of low free energy (low concentration).

Diffusion coefficient (Ds) :

The proportionality constant that measures how easily a specific substances moves through a particular medium. The diffusion coefficient is a characteristic of the substance and depends on the medium.

Diffusion potential :

- A potential that develops as a result of diffusion.
- The potential (voltage) difference that develops across a semipermeable membrane as a result of the differential permeability of solutes with opposite charges (for example K^+ and Cl^-).

Diffusion pressure deficit : Difference between osomotic pressure and turgor pressure and can be expressed as follows :

DPD = OP - TP, OP = TP + DPD and TP = OP - DPD

Diffusion shell : Zone of nutrient depletion around individual roots caused by active nutrient uptake at the root surface and diffusion to the root from the surrounding soil (see also boundary layer).

Diffusional resistance : Restriction posed by the boundary layer and the stomata to the free diffusion of gases from and into the leaf.

Diglyceride : Two hydroxyl groups of glycerol combines with fatty acids.

Dimpled seeds : Uneven shrinking of proteinaceous seeds caused by repeated wetting and drying.

Dioecious :

- Crops in which male and female flowers are present in different plants. E.g. kiwi, papaya.
- Refers to plants in which male and female flowers are found on different individuals (e.g., spinach [*Spinacia*] and hemp [Cannabis sativa]). Contrast with monoecious.

Dioxygenases : Group of oxygenase enzymes that incorporate the two oxygen atoms from O_2 into one or two-carbon compounds.

Diploid : Chromosome no. (2 n)

Direct embryogenesis : Initiation of embryos directly from the explant in absence of callus proliferation.

Direct vigour tests : Vigour tests in which seeds are subjected to environmental stresses found in the field for measuring seed vigour. e.g. Brick grit test, Paper piercing test, Cold test etc.

Direct-inhibition model : Hypothesis for apical dominance stating that the growth of lateral buds is directly inhibited by the relatively high auxin concentration in the stem produced by the shoot apex.

Discolouration of seeds : Variation in seed coat colour caused by ageing or fungal infection.

Disinfestation : Seed treatment to destroy surface borne pathogen/ pest.

Dispersal mode : Means of dispersal observed in plants.

Dispersal unit : A seed or fruit or any propagule that is dispersed by a plant. Syn. Disseminule.

Distortion : Deformity or twisting.

Disulfide bond : Covalent linkage between two sulfhydryl groups on cysteines.

Diterpenes : Terpenes having twenty carbons, four five carbon isoprene units.

Divergent evolution : Naturally selected changes in related species that once shared a common characteristic, but have come to be different during the course of their evolution.

D-Mannitol test : A screening test to measure the drought hardiness of crops using osmoticant, D-mannitol.

DNA denaturation step : Heating of reaction mixture at 90-98^0 C contains the DNA from which a segment is to be amplified, an excess of the two primer molecules, the four deoxyriboside triphosphates and the DNA polymerase.

Dormancy :

- A living condition in which growth does not occur under conditions that are normally favourable to growth.
- Physiological state in which a viable seed fails to germinate when provided with water and environmental conditions normally favourable to germination.
- State of seeds or buds that fail to grow when exposed to an environment that would otherwise have favored germination or growth.

Dormancy (chemical) : Dormancy caused by chemical inhibitory substances in the fruit or seed. e.g. Coriander. Dormancy may be overcome by leaching with water.

Dormancy (combined) : Dormancy as a result of two primary factors, such as seed-coat dormancy and embryo dormancy. Both types of dormancy must be broken for germination to proceed. e.g. scarification followed by prechilling. Syn. Double dormancy.

Dormancy (embryo) : Dormancy as a result of conditions within the embryo itself. e.g. inhibiting substances or incomepletely developed embryo. syn. Endogenous dormancy.

Dormancy (exogenous) : Dormancy caused by the outer structures of the seed or fruit. E.g. impermeability of the seed-coat or pericarp.

Dormancy (mechanical) : Dormancy caused by mechanical resistance of the seed covering (normally the endocarp) to expansion of the embryo. Dormancy is overcome by physically breaking the restricting covering or extracting the seed. e.g. some rice varieties.

Dormancy (photo) : Secondary dormancy developed in light-sensitive seeds. Dormancy is caused by high level of phytochrome Pr that must be changed to phytochrome Pfr by exposure to red light (660-760 nm) or exposed to full day light after imbibitions. e.g. lettuce.

Dormancy (physical) : Dormancy caused by an impermeable seed coat (hard seed). The embryo is quiescent (non-dormant) but is sealed inside the impermeable covering at low moisture content. e.g. Most plants Leguminosae. Scarification can be used to improve germination.

Dormancy (physiological) : A type of embryo dormancy in which germination is prevented by a physiological inhibiting mechanism. e.g. chemical dormancy or thermo-dormancy.

Dormancy (primary) : Dormancy that exists within the seed at the time of maturity on the plant or immediately afterwards.

Dormancy (secondary) : Dormancy that develops within the moist seed after it has been removed from the plant if subjected to adverse environmental conditions. Syn. Induced dormancy.

Dormancy (seed-coat) : Dormancy as a result of seed coat conditions. This type of dormancy may be overcome by scarification or complete removal of the seed coat. e.g. impermeability (physical dormancy) or phytochrome system (photo-dormancy).

Dormancy (thermo) : Primary or secondary dormancy in which the seed must be subjected to a low or fluctuating temperature prior to germination. In temperate regions, thermodormancy is overcome by pre-chilling or stratification, in tropical regions by exposing the seeds to normal diurnal fluctuations in temperature prior to germination.

Dormancy due to embryo : Dormant, immature embryo requires exposure to light for germination Examples lettuce, spruce.

Dormancy due to seed coat : Seed coat impermiable to water, O_2 and mechanically resistant seed coat.

Dorsiventral : Having structurally different upper and lower surfaces.

Double deep dormancy : Dormancy in both the radicle and the epicotyl of the embryo. It may require warm stratification or a combination of warm, followed by cold stratification or cold stratification alone.

Double fertilization : Union of one sperm cell with the egg to form single cell zygote and another sperm unites with two polar nuclei to form triploid endosperm nucleus.

Double sigmoid pattern : Two rapid growth periods separated by period of little growth.

Dough stage : Stage in development of cereal grains when the interior of the kernel (seed) is of dough like consistency.

Down-regulation : Decrease of the normal rate of a process, sometimes involving suppression of genes encoding enzymes involved in that process.

Drip culture : Nutrient solutions are dripped over solid surface.

Drought escape :

- Capacity of a plant to grow and complete its life cycle during the wet season, before the onset of drought.
- Comprises plant that complete their life cycles during the wet season, before the onset of drought.

Drought resistance : Plants capacity to limit and control the consequences of water deficit. Mechanisms include desiccation postponement and desiccation tolerance.

Dry heat :

- Heat applied by dry air as opposed to moist heat, which is applied by moist air.
- Heat provided by dry air that passes through seeds while drying as opposed to moist heat that is provided by moist air.

Dry seed heating : increase in temperature of stored seeds in places of insect activity.

Dry weight : Weight of desiccated (dried) tissue. Often used to measure growth. Avoids variation in water content when fresh weight measurements are made.

DTT (dithiothreitol) : An inhibitor of the enzyme that converts violaxanthin to zeaxanthin. It inhibits blue light-stimulated stomatal opening. A reducing agent that reduces S-S bonds to -SH groups.

DUS : Distinctiveness, Uniformity and Stability. A system of varietal release followed in Western Countries, where a variety or hybrid is released based on its distinctiveness from the existing cultivar, uniformity of its character and stability of its character over a wide range of environmental conditions. This helps in maintaining the genetic purity by the Certification Agency. This test is now also mandatory for registration of varieties under PPV&FR Act, 2001 of India.

Dynamic photoinhibition : Photoinhibition of photosynthesis in which guantum efficiency decreases but the maximum photosynthetic rate remains unchanged. Occurs in moderate, not high, excess light.

❑❑❑

E

Ear cokele : Wheat seed completely converted into nematode (Anguina tritici) gail.

Ecophysiology : Study of the physiological mechanisms by which organisms cope with their environment.

Ecosystem :

- Ecological system that consists of all the organisms in an area and the physical environment with which they interact.
- System involving all ecological or environmental factors, such as climate, soil, physiography, and associated organisms.

Ecosystem nutrient-use efficiency : Is the ratio of litterfall mass to litterfall nutrient

Ecosystem respiration : Sum of plant and heterotrophic Respiration.

Ecotone : Environmental gradient.

Ecotype : Genetically differentiated population that is restricted to a specific habitat.

Ectomycorrhiza : Mycorrhizal association in some trees in which a large part of the fungal tissue is found outside the root.

Ectotrophic mycorrhizal fungi :

- A dense sheath of fungal mycelium around the roots and extending into the surrounding soil; some hyphae may penetrate between, but not into, the root cortical cells.
- Thick sheath or mantle of fungal mycelium around the roots, and some of the mycelium penetrates between the control cells.

Edaphic factor : The physical, chemical or biological characteristic of soil that affects plant growth and seed production effect.

Efficiency : Rate of a process per unit plant Resource.

EGTA (ethylene glycol-bis (β-) aminoethyl ether- N, N, N′, N′- tetra acetic acid) : A compound that chelates calcium ions, preventing their uptake by cells and inhibiting both root gravitropism and the asymmetric distribution of auxin in response to gravity.

Eicosanoids : A group of substances involved in the mammalian inflammatory response and similar to jasmonic acid found in plants.

EIN mutants : In Arabidopsis, mutants blocked in their ethylene responses (ethylene-insensitive).

EIN2 gene : In Arabidopsis, ethylene response gene encoding a protein that may act as a membrane channel protein.

EIN3 gene : In Arabidopsis, ethylene response gene encoding a transcription factor.

Elastic biological strain : An organisms function returns to optimum level when conditions are again optimum.

Elastic modulus : Force needed to achieve a certain reversible change in cell volume.

Electric hygrometer : Electric instrument for measuring the degree of moisture in the atmosphere.

Electric moisture meter : An instrument that measures seed moisture based on the di-electric property of seeds where the resistance offered by the seed is inversely proportional to moisture content. e.g. Universal moisture meter.

Electro fusion : Fusion of protoplasts induced by electric shock.

Electrochemical potential : The chemical potential of an electrically charged solute.

Electrochemical potential gradient : Electrical potential of a substance considering the importance of electric fields in biological transport.

Electrogenic pump : Active transport mechanism due to movement of net electric charge happens when an ion moves into or out of a cell without being balanced by counter movement of an ion of opposite charge, a voltage is created across the membrane.

Electrogenic transport :

- Active ion transport involving the net movement of charge across a membrane.
- Ion transport involving the net movement of charge across the membrane.

Electrolytes :

- Molecules dissociate in to ions when dissolved. Example - NaCl which dissociates in to Na^+ and Cl^- ions.
- Substances which dissociate into ions on dissociation.

Electron spin resonance (ESR) : A magnetic resonance technique that detects unpaired electrons in molecules. Instrumental measurements that identify intermediate electron carriers in the photosynthetic electron transport.

Electron transport chain : In oxidative phosphorylation, electrons are transferred along an electron transport chain consisting of a collection of electron transport proteins bound to the inner of the two mitochondrial membranes. System transfers electrons from NADH produced in glycolysis, kreb's cycle or pentose phosphate pathway to oxygen.

Electron transport chain (in the mitochondrion) : A series of protein complexes in the inner mitochondrial membrane linked by the mobile electron carriers ubiquinone and cytochrome c, that catalyze the transfer of electrons from NADH to O_2. in the process a large amount of free energy is released. Some of that energy is conserved as an electrochemical proton gradient.

Electronegative : Having a negative electric charge.

Electroneutral transport :

- Active ion transport that involves no net movement of charge across a membrane.
- Involves no net movement of charge across the membrane. Example - Na^+ / K^+ - AT pase of animal cells pumps, three Na^+ ions out for every two K^+ ions in, resulting in a net outward movement of one positive charge.

Electronic colour sorter : This machine separate off-coloured and discoloured seed from normal coloured ones. It uses photoelectric cell that compares the seed coat colour to a selected background and reject all those seeds that do not match.

Electrophoresis :

- Migration of colloidal particles to different charged poles due to electric charges.
- The movement of protein molecules suspended in solution (polyacrylamide gel, agarose gel, urea) towards electric poles.

Electrophoresis pattern : The pattern of protein bands subjected to an electric field used for varietal identification.

Electrophoresis test : A method for separating and mapping of protein bands from homogenized plant seed extracts in gel preparation across an electric field. This is used in varietal identification.

Electroporation : Temporary opening in the membranes of protoplasts by providing short electric pulses with the generator to facilitate uptake of DNA.

Elevator : Machine that helps in upward movement of seed in a processing unit. E.g. Bucket elevator.

Elf3 mutant : In Arabidopsis, a flowering-time mutant (flowering 3).

Elicitors :

- The specific pathogen molecules recognised by the plant.
- Specific pathogen molecules or cell wall fragments that bind to plant proteins and thereby signal for plant defense against a pathogen.

ELISA : Enzyme-linked immunosorbent assay. A very sensitive detection method that uses radioisotopes or chemiluminescent compounds linked to antibodies to detect compounds (like IAA) or proteins in tissue extracts.

Elongation : Growth of the plant axis or cell primarily in the longitudinal direction.

Elongation zone (of a root) :

- The region of rapid and extensive root cell elongation showing few, if any, cell divisions.
- Begins 0.7 to 1.5 mm from the apex.

Embryo :

- A basic body plan of the mature plant and many of the tissue types of the adult though these are present in a rudimentary form.
- Immature plant formed after sexual or asexual reproduction. Found in seed (of seed plants) and consists of an embryonic axis bearing a terminal bud (plumule), root (radicle), and one or more seed leaves (cotyledons).
- Part of seed consists of plumule and radicle.
- Rudimentary plant within the seed.
- Seed embryo is a rudimentary plant consisting of primary axis and cotyledon.

Embryo axis : Hypocotyl-root axis bearing at one end the root meristem and at the other the cotyledon or cotyledons and the shoot meristem.

Embryo culture : Culture of isolated mature or immature embryos.

Embryo dormancy : Dormancy as a result of conditions within the embryo either due to immaturity or inhibitors. Syn. Internal dormancy.

Embryo rescue : Rescue of hybrid embryos from the maternal environment following certain wide (intergeneric) sexual crosses to prevent embryo abortion in early developmental stages.

Embryo sac :

- Embryo carrying body.
- A thin walled sac embedded in the nucellus and contains the egg.
- Centre part of ovule where embryogenesis occurs.
- In flowering plants, large oval cell (arising from the megaspore) in the ovule that develops into the female gametophyte and in which fertilization of the egg and development of the embryo take place.

Embryogenesis :

- The development of a zygote in to a multicellular embryo. In plants, the processes of cell division and differentiation that take place in the ovule and immature seed and establishes the basic developmental patterns of the adult plant : the radial pattern of tissues; the apical-basal axis; and the primary meristems.
- The developmental process which initiated plant development. The process transforms a single celled zygote into a multicellular microscopic embryonic plant.
- Origin of plantlets by first developing the embryos from a group of cells asexually as somatic embryogenesis or from a fertilized egg cell as sexual embryogenesis.

Embryogenic determined cells (IEDC) : Cells from which embryo arise.

Embryoid : Plantlet, embryolike in structure, produced by somatic cells in vitro, also adventitious embryos developing in vitro by vegetative means.

Embryonic axis : The hypothetical central line of an embryo around which lateral organs are arranged.

Emerson effect : Enhancement of photosynthetic rate under influence of two wavelengths (long and short).

emf mutant : In Arabidopsis, a mutant of EMBRYONIC FLOWER (EMF) that produces flowers soon after germination; the shoot apical meristem initiates no vegetative structures, but immediately initiates floral organs during germination.

Emissivity : Coefficient that describes the thermal radiation emitted by a body at a particular temperature relative to the radiation emitted by an ideal black body.

Empty seeds-purity analysis : In purity analysis empty seeds belonging to families *Cucurbitaceae, Solanaceae* without embryo and endosperm are regarded as pure seeds.

Emulsion : Both the phases are liquid. Examples - Milk cream, latex of calotropis.

Encrusted seed : Seed units more or less retaining the original shape of seed after pelleting even though the size and weight are changed.

End product inhibition : Decline in photosynthetic rate in leaves due to sugar accumulation through photosynthesis. Sugars tieing up with Pi (inorganic phosphate) results in Pi limitation of RUBP regeneration capacity.

Endocytosis : Uptake of material into a cell by an invagination of the plasma membrane and its internalization in a membrane-bound vesicle.

Endodermis :

- Layer of thick wall cells due to deposition of lignin, suberin and cutin adjacent to cortex.
- A specialized layer of cells with a Casparian strip; surrounding the vascular tissue in roots and some stems.
- Central ring of cells adjacent to cortex.
- Innermost layer of root cortical cells that surrounds the vascular tissue; these cells are surrounded by a suberized Casparian strip that blocks apoplastic transport.

Endogenous : Relating to the interior of a living system. Concerning what is found within or originates within such a system.

Endogenous oscillator : An internal, molecular pacemaker that maintains circadian rhythms in a constant light or dark environment.

Endogenous rhythm : A rhythm that persists in the absence of external controlling factors such as light.

Endoplasmic reticulum : Network of internal membranes which are typical lipid bilayers with interspread integral and peripheral proteins.

Endoreduplication : Doubling of chromosomes in mitosis with lack of subsequent cytokinesis.

Endosperm :

- Produced by fusion of one sperm nucleus with two polar nuclei, containing 3n chromosome set, associated with storage of food material.
- Tissue of seeds, developing from fertilization of the polar nuclei of the ovule by the second male nucleus; it nourishes the embryo.

Endosymbiosis : A theory that explains the evolutionary origin of the chloroplast and mitochondrion through formation of a symbiotic relationship between a prokaryotic cell and a simple nonphotosynthetic eukaryotic cell, followed by extensive gene transfer to the nucleus.

Endosymbiotic origin : A theory in which the engulfment of a free-living ancestral prokaryotic cell by a eukaryotic host evolved into a cell with semi-autonomous chloroplasts and mitochondria.

Energy constant : Magnitude of energy a photon contains expressed as follows $E = hc/\beta$

where E = Energy constant,

h = plank's constant (6.625×10^{-27} erg/sec).

C = velocity of light (3×10^{10} cm/sec). β = Wave length size.

Energy transfer : In the light reactions of photosynthesis, the direct transfer of energy from an excited molecule, such as carotene, to another molecule, such as chlorophyll. Energy transfer can also take place between chemically identical molecules, such as chlorophyll-to-chlorophyll transfer.

Energy utilized :

$$\text{Hayashi (1969) Eu} = \frac{\text{Fixed energy}}{\text{Intecepted energy}}$$

Fixed energy : 1 g dry weight = 4000 Cal.
(Hayashi, 1969; Varshmy, 1972).

Enhancement effect : The synergistic (higher) effect of red and far-red light on the rate of photosynthesis, as compared with the sum of the rates when the two different wavelengths are delivered separately.

Entrainment :

- Synchronizing effects of light at day break in nature which brings their periods closer to 24 hours.
- The synchronizing effects of external controlling factors, such as light and darkness, on the period of biological rhythms.

Envelope : the double membrane system surrounding the chloroplast or the nucleus. The outer membrane of the nuclear envelope is continuous with the endoplasmic reticulum.

Ephemeral : Short lived.

Epiblema : Outermost layer of root possess root hairs.

Epicotyl :

- Part of seedling axis above cotyledons.
- The part of the seedling axis immediately above the cotyledons and below the primary leaf.
- The region of the seedling stem above the cotyledons.

Epidermis (epidermal cells) :

- The outermost layer of plant cells, typically one cell thick.
- Outermost cell layer of an organ, typically covered by a cuticle.

Epigeal germination :

- A mode of germination in which cotyledons are carried above the soil level by the elongation of hypocotyls.

- Germination process in which cotyledons come above soil (Epi- upon, geal - earth). Examples - Cotton, Papaya, Sem, Castor, Sunflower.

Epigenetic regulation : Changes in gene expression that are stable even after the signal that induced the change (in this case cold) is removed .

Epinasty :

- A downward curvature of leaves due to asymmetric growth of the petiole. A response to ethylene production during flooding.
- Downward bending of a plant organ.
- Downward curvature of leaves that occurs when the upper (adaxial) side of the petiole grows faster than the lower (abaxial) side.

Epiphyte : Plant living on another plant as a support, without a symbiotic or parasitic association.

Epstein (1972) : Element essential for plant life cycle with a part of essential plant molecule.

Equilibrium : For a particular solute, a state or condition in which there is no gradient of (electro)- chemical potential and therefore no net passive transport of the solute.

Equilibrium moisture content : Moisture content of seed attains equilibrium with the surrounding atmosphere at a given relative humidity, temperature and atmospheric pressure.

Equivalent heat : An exposure of 160°C dry heat for 12 hours and sterilization (steam) at 121°C for 15 minutes.

ERE- Binding proteins (EREBPs) Proteins that bind to ERE.

Escape from photoreversibillity : The loss of photoreversibility by farred light of phytochrome mediated red light induced events after a short period of time.

Essential element : A chemical element that is part of a molecule that is an intrinsic component of the structure or metabolism of a plant. When the element is in limited supply, a plant suffers abnormal growth, development, or production.

Essential oils : Mixtures of volatile terpenes and other secondary metabolites that give characteristic odors to some plants, e.g., peppermint, lemon, basil, and sage.

Ethephon : An ethylene releasing compound, 2-chloroethylphosphonc acid, that makes practical the application of the plant hormone ethylene gas under field conditions. Grade name, Ethrel.

Etheral : Chemical which releases ethylene on application.

Ethylene ethene (C_2H_4) :

- Ethylene gas (CH_2=CH_2) functions as a plant hormone that is synthesized from the amino acid methionine via ACC. Has important effects on plant growth and development, including stimulating or inhibiting elongation of stems, roots, depending on conditions and species; enhances fruit development; suppresses flowering in most species; increases abscission of flowers and fruits. Increases RNA transcription of numerous genes.
- Gaseous hormone required in dormancy breaking, epinasty and artificial fruit ripening.
- A gaseous phytohormone; ethylene is, e.g., a signaling compound when roots are exposed to hypoxia, inducing aerenchyma formation and petiole extension.

Ethylene response element : Key regulatory sequence found in genes regulated by ethylene.

Ethylene response factors (ERF1) : Which incodes a protein that belongs to the ERE - binding protein (EREBP) family of transcription factors which were first identified in Tobacco as proteins that bind of ERE sequences (Ohme Takagi and Shinshi 1995).

Ethylene triple response : Common responses to ethylene of growing etiolated seedlings of most dicots and of coleoptiles and mesocotyls of grass seedlings (e.g., wheat and oats). Reduced rate of elongation, increased lateral expansion, and swelling in the region below the hook.

Ethylene-insensitive mutants : Mutants that do not respond to ethylene. In *Arabidopsis,* tall seedlings protruding above short seedlings showing ethylene triple response when grown in the presence of ethylene.

Etiolated growth : Pale, unusually tall and spindly appearance of seedlings grown in dark.

Etiolated seedlings : Dark grown seedlings in which the hypocotyl and stem are more elongated, cotyledons and leaves do not expand, and chloroplasts do not mature.

Etiolation :

- The form and growth of seedlings grown in darkness. A pale, unusually tall and slender appearance, dramatically different from the stockier, green appearance of seedlings grown in the light.
- Weak, long and thin seedling with minute yellow leaves.

Etioplasts :

- Photosynthetically inactive form of chloroplast found in etiolated seedlings. Does not synthesize chlorophyll or most of the enzymes and structural proteins required for the formation of thylakoids and operation of photosynthesis. Contains an elaborate system of interconnected membrane tubules called the prolamellar body.
- Formed from conversion of proplastids when seeds are germinated in the dark.

Etr1 mutant : In *Arabidopsis*, a dominant mutation that blocks the response to ethylene, (ethylene resistant 1).

Evaporative heat loss : Loss of heat due to the cooling resulting from the evaporation of water.

Evapotranspiration :

- Transpiration coupled with evaporation.
- Water loss from an ecosystem by transpiration and surface evaporation.

Evergreen : Bearing foliage that persists and remains throughout the year, as opposed to deciduous.

Exalbuminous : Seeds that do not possess an endosperm. Examples : Cycocel, B-Nine, phosphon D etc.

Exanthema disease : Secretion of gummy exudates due to lack of copper.

Excised embryo test : Used for those seeds that normally germinate slowly (tree seeds) or show dormancy. The embryos of such seeds are removed and tested for growth on paper medium.

Exclusion : Prevention of net entry of a molecule; it may be due to low permeability for a molecule or to its extrusion.

Excretion : Active secretion of compounds (e.g., salt from leaves).

Exhaustion test : A vigour test to differentiate vigour in seeds. Seeds are placed in between germination paper 5 cm below the margin and kept in dark. Those seeds that have enough vigour will grow above the marker line while low vigour seeds will fall.

Exodermis (or hypodermis) :

- In mature roots, an outer layer of cells of the cortex that is relatively impermeable to water.
- Contain hydrophobic materials in its walls and is relatively permiable to water.
- Outer cortical cell layer in roots, immediately below the epidermis; these cells are surrounded by a suberized Casparian strip that blocks apoplastic transport.

Exogenous : Relating to or coming from outside a living system. Originating outside such a system.

Expansins :

- Class of wall loosening proteins that accelerate wall stress relaxation and cell expansion, typically with an optimum at acidic pH. Appears to mediate acid growth.
- Wall loosening proteins found in the cell wall of wide range of plant species.
- Cell-wall enzyme involved in cell Expansion.

Expansion : Cell and tissue growth due to increase in cell size, not cell number.

Export : The movement of photosynthate in sieve elements away from the source tissue.

Extensin : Rigid cell-wall glycoprotein, rich in hydroxyproline, that represents 5–10% of the dry weight of most primary cell walls; significant component of the secondary walls of sclerenchyma cells.

Extinction coefficient :

- Coefficient describing the exponential decrease in irradiance through a compartment that absorbs radiation (e.g., a leaf, a canopy, or a pigment in solution).
- Ratio of light intensity at the middle of canopy (I_m) to the incident upon plant (I_o)

$$Ec = \frac{I_m}{I_o} \times 100$$

Extractive biotechnology : Use of substances that are produced naturally. Examples - Sugarcane, oils, fats, dye, pharmaceutically active materials.

Extrusion : Ion transport from root cells to the external medium, dependent on respiratory metabolism.

Exudate : Compounds released by plants (mostly by roots); also xylem or phloem fluid that appears when the stem is severed from the roots or a cut is made in the stem.

Exudation : Release of exudates, or the appearance of fluid from cut roots or stems.

❑❑❑

F_0F_1- ATP synthase : A multi-protein complex associated with the inner mitochondrial membrane that couples the passage of protons across the membrane to the synthesis of ATP from ADP and phosphate. The subscript 'o' in Fo refers to the binding of the inhibitor oilgomycin. Similar to CF_0-CF_1 ATP synthase in photophosphorylation.

Facilitated diffusion : Passive transport across a membrane using a carrier.

Facilitation : Positive effect of one plant on Another.

Facultative anaerobes : Requires O_2 for respiration but can respire anaerobically.

Facultative CAM plants : Plants that photosynthesize by Crassulacean Acid Metabolism (CAM) during dry periods and by C_3 or C_4 photosynthesis at other times.

Far red light : Light wave spectrum in the range of 710-720 nm that inhibits germination.

Farm saved seeds : Seeds used by the farmer from the crop harvested and stored on his farm.

Fast-green test : A method of assessing mechanical damage on seeds by soaking it in 0.1% fast green solution. Upon washing green streaks deposited on the crack that will reveal the damage to the naked eye.

Fatsor Lipids : Esters of fatty acid and glycerol (Triphydric alcohal and free fatty acid).

Fatty acids :

- Long chain aliphitic acid.
- Propectin and pectic acid are closely related to hemicellulose and are polymer uronic acid.

Feedback : Influence of a product of a later step in a chain on an earlier step; fluctuations in rate of the process or concentration of metabolites are minimized with negative feedbacks or amplified with positive feedbacks.

Feeder layer : Layer consists of X irriated nondividing but living protoplasts plated in sugar medium in petridishes.

Feedforward : Response in which the rate of a process is affected before any deleterious effect of that process has occurred; for example, the decline in stomatal conductance before the water potential in leaf cells has been affected.

Female gametophyte : Body carrying female gamete.

Fermentation :

- Pyruvic acid obtained from glycolysis yields ethyl alcohal or lactic acid.
- A process of seed extraction in case of fleshy fruits by the action of bacteria, yeast or fungi. Fruits are crushed and added with water. Bacteria proliferate and release the seed from mucilage. **e.g.** Cucumber, tomato.
- Anaerobic conversion of glucose to organic acids or alcohol.

Fermentative metabolism : The metabolism of pyruvate in the absence of oxygen, leading to the oxidation of the NADH generated in glycolysis to NAD^+.

Ferredoxin (Fd) : A small, water soluble, iron sulphur protein involved in electron transport in photosystem I.

Ferredoxin-thioredoxin system : Three chloroplast protein (ferredoxin, ferredoxin-thioredoxin reductase, thioredoxin). The concerted action of the three proteins uses reducing power from the photosynthetic electron transport system to reduce protein disulfide bonds by a cascade of thioldisulfide exchanges. As a result, light controls the activity of several enzymes of the Calvin cycle.

Ferric chloride test : A test to determine the extent of mechanical damage on seeds. Seeds are soaked in 20% ferric chloride solution for 5-15 minutes and then washed in water to reveal the dark streaks embedded on cracks.

Ferrous sulphate test : A test for varietal identification of sorghum cultivars. The seeds are soaked in one percent ferrous sulphate solution for four hours at 35°C and later classified based on colour changes.

Fiber : An elongated tapered sclerenchyma cell that provides support in vascular plants.

Fibrous root system : The complex root system of monocots, which lack a main root axis; all roots have about the same diameter.

fick's first law : The rate of diffusion is directly proportional to the concentration gradient, defined as the difference in concentration of a substance between two points separated by the distance ?÷.

Field capacity or water holding capacity :

- Water soil retains against gravitational force after the excess of water is removed or moisture holding capacity of soil.
- The water content of a soil after it has been saturated with water and excess water has been allowed to drain away. The moisture-holding capacity of soils.
- Water content that a soil can hold against the force of gravity.

Field count : The prescribed procedure to inspect a seed field by observing a sample of plants in the seed crop to assess the contamination.

Field drying : Drying of seed on the mother plant.

Field emergence :

- Emergence of seedlings from seeds planted in the field; usually refers to percentage of seeds planted in the field that germinates and produces seedlings strong enough to emerge through the soil.
- The percentage of seeds emerged in the field to the total seed planted.

Field fungi : Molds that live outside on growing crops.

Field weathering : Damage to the matured seeds retained on mother plant caused by weather factors like rain, wind, frost, etc.

Film coating : Seeds are coated using polymer based chemical additives as a thin film so that shape of the seed remains unchanged. e.g. Polyethylene-oxide, Polyvinylalcohol, Polyvinyl-pyrolidone.

Filter sterilization : Use of bacteria proof membrane filters viz. millipore, Nalgene, sartorius for sterilizing growth factors, vitamins, toxins etc.

Final count : The day on which the germination count is taken as prescribed by ISTA rules; the normal seedlings are counted and reported as germination percentage.

Fine grinding : A grinding specification of cereals and oil seeds for moisture estimation in which at least 50% of the ground material should pass through 0.5 mm sieve.

First count :

- Number of normal seedlings counted during first count (4/ 5^{th}day), represents faster germinating seeds indicating seed vigor.
- During conduct of germination test, the first count is done during the middle of the prescribed final count. This is done to assess the vigour (speed of germination) and to remove disease seeds.

First excited state : Excited state of chlorophyll molecule after light absorption.

Fish mouth : A type of imbibition injury in soybean seeds due to differential absorption and desorption of seed coat resulting in crack of seed coat resembling a fish mouth.

Flame sterilization : Soaking the instruments in 70-80% alcohal followed by flaming on a burner in a laminar air flow hood.

Flavanols, flavines, flavones : Families of Flavonoids.

Flavin adenine dinucleotide (FAD) : A riboflavin-containing cofactor that undergoes a reversible one or two electron reduction to produce FADH or $FADH_2$

Flavin hypothesis : Discredited suggestion that riboflavin activated by blue light could participate in the in vivo photodestruction of an auxin. Although blue light can reduce a flavin such as riboflavin in vitro, and the reduced flavin can in turn reduce cytochrome c, no in vivo role for these photoreactions has been shown.

Flavin mononucleotide (FMN) : A riboflavin-containing cofactor that undergoes a reversible one or two electron reduction to produce FMNH OR $FMNH_2$.

Flavones and Flavonols :

- Other major group of flavonoids absorb light of shorter wavelengths and not visible by human eye. However, insects such as bees which see further into the ultraviolet range of the spectrum may respond to flavones and flavolons as attractant cues. They protect cells from excessive UV-B radiation (280-320 nm) because they absorb such wavelengths while allowing photosynthetically active wavelengths to pass through uninterrupted.
- Croup of ultraviolet light-absorbing, protective flavonoids that may also attract pollinating insects to flowers. Secreted along with flavonoids into the soil by legume roots, they mediate interaction with nitrogen-fixing symbionts.
- Group of ultraviolet light-absorbing, protective flavonoids that may also attract pollinating insects to flowers secreted into the soil by legume roots, they mediate interactions with nitrogen fixing symbionts.

Flavonoids :

- Derivations of phynyl propane found in chromatophores, vacuoles provided colour to the plant or plant part.

- One of the largest classes of plant phenolics, in which two aromatic rings are connected by a carbon link to a third phenyl ring; representatives of this class play a role in the symbiosis between rhizobia and legumes, as phytoalexins, as antioxidants, in the colors of flowers and as defense compounds.

- One of the largest classes of plant phenolics contain 15 carbons arranged in two aromatic rings connected by a three carbon bridge. The structure results from two seperate biosynthetic pathways Shikimic acid pathway and malonic acid pathway.

- A large group of plant phenolics with the basic carbon structure of two aromatic rings connected by three carbons which includes the anthocyanins, the flavones, the flavonols, and the isoflavones. Functions in plant pigmentation, protection against UV irradiation, and defense against herbivores and pathogens.

Flavoprotein ferredoxin-NADP reductase (Fp) : In photosystem I, a membrane-associated flavoprotein that reduces $NADP^+$, using electrons from ferredoxin. This reaction completes the noncyclic electron transport that begins with the oxidation of water.

Floaters : Seeds that float on water or salt solution due to density difference; these are mostly ill filled, immature or insect damaged.

Flooding sensitive plants : Plants which are severely damaged by 24 hours of anoxia. Examples garden pea ***(Pisum Sativum)***

Flooding tolerant plants : Plants which can withstand anoxia (lack of oxygen).

Flooding-sensitive : Refers to plants that are severely damaged by 24 hours of anoxia due to flooding of roots.

Flooding-tolerant : Refers to plants that can withstand anoxia temporarily, but not for more than a few days.

Floral differentiation : Development of floral part viz; calyx, corolla, stamens and carpels.

Floral evocation : The events occurring in the shoot apex that specifically commit the apical meristem to produce flowers.

Floral homeotic genes : Key regulatory genes that determine the positions and identities of floral organs in flowers.

Floral homeotic mutants : The genes that determine floral organ identity.

Floral meristem :

- Forms floral (reproductive) organs : sepals, petals, stamens, and carpels. May form directly from vegetative meristems or indirectly via an inflorescence meristem.
- Meristems which produce floral organs.

Floral organ identity genes : Genes which directly control floral identity. The proteins encoded by these genes are transcription factors that likely control the expression of other genes whose products are involved in the formation and or function of floral organs.

Floral organs : Angiosperm organs involved directly or indirectly in sexual reproduction; sepals, petals, stamens, and carpels.

Floral primordia : A primordium is the earliest recognizable stage when a group of cells begins to form a structure. In the developing Arabidopsis flower, four discrete whorls of floral primordia are formed from a floral meristem, and these primordia give rise to sepals, petals, stamens, ovules, style, and stigma.

Floral reversion : The conversion of a floral meristem to a vegetative or inflorescence meristem, causing a shoot or inflorescence to grow directly out of the developing flower.

Floral stimulus : In photoperiodism, signals that are translocated from the leaves to the shoot apical meristem.

Florigen :

- Hormone causing flower intiation.
- The hypothetical, universal flowering hormone synthesized by leaves and translocated to the shoot apical meristem via the phloem. So far, it has not been isolated or characterized.
- In the 1930 S, Mikhail Chailakhyan, working in Russia, postulated the existance of a universal flowering hormone , which he named florigen.

Flower : Specialized reproductive shoot structure of angiosperms. Consists of non-reproductive organs (sepals and petals) and reproductive organs (stamens and carpels).

Flower induction : Formation of floral primordia by shoot meristem instead of leaves.

Fluence :

- Number of photons impinging on a unit surface.
- Response response to a dosage of light.
- The number of photons absorbed per unit surface area.

Fluence rate : A unit for the measurement of light falling on a spherical sensor from many directions expressed as watts per square meter (W m^{-2}) or moles of photons per square meter per second (mol m^{-2} s^{-1}).

Fluorescein diacetate (FDA) : Used to tetect protoplast viability. As FDA accumulates in the cell membrance, viable intact protoplasts fluoresce yellow green within 5 minutes used at concentration of .01% dissolved in acetone.

Fluorescence :

- Loss of energy as heat during return of electrons of chlorophyll from the higher energy levels to the ground state.
- Following light absorption, the emission of light at a slightly longer wavelength (lower energy) than the wavelength of the absorbed light.
- Photons emitted when excited electrons return to the ground state.

Fluorescent labeling : Isolation of fusion products from parental protoplasts by loading with fluorescent dyes prior to fusion.

Flux : The rate of flow of a substance expressed in units of quantity through a unit surface area per unit time.

Flux density (J_s) :

- The rate of transport of a substance across a unit area per unit time. Js may have units of moles per square meter per second [mol $m^{-2}s^{-1}$].
- Amount of substances crossing a unit area per unit time expressed as mol. m^{-2} S^{-1}.

Foliar application or aerial spray :

- Technique of spraying the mineral nutrients to the leaves.
- The application and absorption of some mineral nutrients to leaves as sprays.

Foot candle : An illuminance of one lumen per square foot. (10.76 lux=1ft.c.).

Formative divisions : Cell divisions that form new cell files in an axis, as in a root apical meristem, usually longitudinally oriented.

Formazan : Red coloured compound formed due to reduction of 2, 3, 5 triphenyl tetra zolium chloride in living cells due to action of dehydrogenase enzymes.

Forst injury : Freezing temperature at higher altitude damage developing seeds due to formation of ice crystals in the cells.

Free energy : A thermodynamic term representing a capacity to perform work.

Free flowing seeds : A term used in processing to describe those seeds that flow freely when dropped from a container. e.g. Wheat, Rice, Beans etc.

Free moisture : Moisture content of seed above the equilibrium moisture content; this can be easily removed from the seed.

Free radical pathology : A theoretical term used to describe the effect of free radical at the cellular level affecting the viability of seeds and eventually leading to death of seeds.

Free radical quenching : A theory postulated by Prof. R.N. Basu of Calcutta University. The aged seeds are soaked in water or dilute solutions of salts for prescribed period (1 to 6 hrs.) so that the unpaired electrons of the fee radicals get the opposite charge and get quenched or become non reactive so that further damage is controlled.

Free-running : Designation of the biological rhythm that is characteristic for a particular organism when environmental signals (zeitgebers) are removed, as in total darkness.

Free running rhythm : Rhythm which occurs without environmental signals.

Free water : A form of water that is held in the seed beyond 20% seed moisture content and can be removed on drying.

Freeze-drying : Process for drying seeds and other materials; it uses a freezing temperature combined with moving dry air, which sublimes frozen water to water vapour.

Freezing injury :

- Injury occurs at temperature below freezing point of water.
- Injury that occurs when plants are cooled below the freezing point of water. Contrast with chilling injury.
- Damage caused to seeds due to exposure to low temperature on account of formation of ice crystals in cells.

Frequency (V) : A unit of measurement that characterizes waves, in particular light energy. The number of wave crests that pass an observer in a given time.

Fresh ungerminated seeds : Seeds that remain firm and viable at the end of germination test even after treatment for breaking dormancy. **Syn.** Swollen seeds.

Fresh weight : Weight of living tissue.

Frost : Hardening acclimation of a plant as a result of exposure to low temperatures that make it frost tolerant (e.g., hardening in autumn).

Frost hardiness/tolerance : Physiological condition that allows exposure to subzero temperatures without cellular damage.

Fruit (s) :

- Ripened (matured) ovary of a plant, together with any intimately attached parts that developed with it from the flower. Fruits of buckwheat, cereals, grasses, sunflower, and many others are commonly called seeds.
- In angiosperms, one or more mature ovaries containing seeds and sometimes adjacent attached parts.

Fruit maturity : Stage of fruit ready for harvesting.

Fruit ripening : Stage of fruit for consumption.

Fruit set :

- Initiation of ovule growth after successful pollination.
- The start of fruit growth following pollination.

Fumigant :

- A chemical substance, which at specified temperature and pressure can exist in gaseous form in sufficient quantities to be lethal to pests.
- Chemical applied as a gas to destroy molds, insects, rodents, etc.

Funicle : Stalk of ovule.

Furanocoumarins : Group of coumarins with attached furan rings whose toxicity may result from exposure to light.

Fusicoccin :

- A fungal phytotoxin which stimulates rapid proton extrusion and transient growth in stem and coleoptile sections.
- A fungal toxin that induces acidification of plant cell walls by activating an H^+- ATPase in the plasma membrane. Fusicoccin stimulates rapid acid growth in stem and coleoptile sections. It also stimulates stomatal opening by stimulating proton pumping at the guard cell plasma membrane.

Fusiform stem cells : Elongated, vacuolate stem cells of the vascular cambium that divide longitudinally and whose derivatives form the conducting cells of the secondary xylem and phloem.

❑❑❑

G Protein :

- GTP-binding protein involved in signal transduction,
- Intracellular membrane-associated proteins activated by several receptors.

GA1 mutants : In Arabidopsis, dwarf mutants deficient in biologically active gibberellins.

GA_{12}-aldehyde : The end-product of the second stage of giibberellin biosynthesis and the precursor for all other gibberellins.

GA2 oxidase : Enzyme inactivates GA1, by catalysing the addition of a hydroxyl group to C-2.

GA3 oxidase : Functions as a 3 B hydroxylase, adding a hydroxyl group to C-3 to form the active gibberellin.

GA20 oxidase : Enzyme catalyses all the reactions involving the successive oxidation steps of carbon 20 between GA 53 and GA 20, including the removal of C-20 as CO_2.

GADA test :

- Amount of carbon dioxide evolved by glutamic acid reflects the level of enzyme glutamic acid decarboxylase which is related with vigour and storability of seed.
- A physiological vigour test that measures the Glutamic Acid Dehydrogenase enzyme activity correlated with vigour.

GAI gene : In Arabidopsis, encodes a transcription factor that represses GA responses in the absence of a GA signal.

Gall : An unorganized mass of tumour like plant tissue resulting from uncontrolled cell division.

Gamet divider : A divider for mixing dividing seeds using centrifugal force. The divider consists of a spinning cup attached to a rod that is rotated by a motor. As the seeds fall down the divider, they are dispersed by the rotating cup and are divided by a ridge placed below. The seeds come through two outlets. **Syn.** Centrifugal divider, Precision divider.

Gas Chromatography (GC) : Method that separates components of a mixture according to their affinity for inert column material versus their tendency to volatilize and move through the column on a current of inert gas.

Gas storage : Storage in an atmosphere composed of a gas other than air.

Gaseous water : Water vapour.

Gate/Gates :

- A structural domain of the channel protein that opens or closes the channel in response to external signals such as voltage changes, hormone binding or light.
- Structures in channel proteins that can open or close the pore in response to external signals. Signals that can open or close gates include voltage charges, hormone **binding or light. For example** : Voltage channels open or close in response to changes in the membrane potential.

Gauge : A term used in UK to define the thickness of thin film like polythene. 100 Gauge (UK) = 25 μ or 0.025 mm; 700 Gauge = 0.175 mm.

G-box : A specific sequence of DNA nucleotides that bind G-box type cis-acting transcription factors, leading to the transcriptional activation of genes.

Gel/Gels :

- A highly-hydrated, dispersed network of long polymers, typically with elastic properties intermediate between that of a liquid and that of a solid.
- Colloids which are more viscous and rigid having the consistency of semisolids. Examples - Table jelly, set custurd.

Gene activators : Proteins that, either alone or in concert with other proteins, increase gene expression.

Gene bank : A large collection of germplasm or plant species from various parts of the world.

Gene li (liguless) : Conditions presence or absence of a ligule at the base of the leaf in barley.

Generative cells : Cell of pollen grain which undergoes mitotic division to form two sperm cells during pollen tube formation at fertilization.

Genetic engineering :

- Transfer of genes from bacteria, virus or other means to the plants to impart resistance against antibiotics and other substances.
- Transfer of specifically constructed gene assemblies through various transformation techniques.

Genetic tumours : Spontaneous tumors produced by certain genotypes. Form in about ten percent of interspecific crosses of the genus. Nicotina due to overproduction of cytokinin.

Genotype : Genetic constitution of an organism.

Geotropism : Growth response of plant organs with respect to gravity.

Geranylgeranyl diphosphate (GGPP) : A precursor in the synthesis of gibberellins.

Germinability : Active growth of the embryo of a seed/Resumption of growth by embryo and developement of young plant from seed.

Germination :

- Emergence and development of seedling with all those essential structures that for the species described indicates the ability to develop into a normal seedling under favourable condition of soil (ISTA, 1985).
- Emergence of radicle from the seed.
- The beginning or resumption of growth by a spore, seed, or bud.

Germination (of a seed) : Emergence of a part of the embryo through the seed coat, normally the radicle.

Germination capacity : Proportion of seed sample that has germinated normally within a specified test period. **Syn.** Germination percent.

Germination energy : A vigour rating of seed germination determined by germination percent of a lot at its peak.

Germination index : Summation of number of seedlings emerging on a specific day divided by the number of day on which the germination was estimated.

Germination inhibitors : Substances which inhibit germination by retarding enzymatic activities required for germination process.

Germination percent : The percentage of seed in a sample that has germinated normally.

Germination value : Is a relative measure of rate of germination and is given by formula (Czabator, 1952)

$$GV = \frac{\text{Peak value of cumulative number of normal seedling}}{\text{Days of germination count}}$$

$$X \frac{\text{Total number of normal seedling}}{\text{Days of final count}}$$

Germplasm : Living substance of the cell nucleus that determines the hereditary properties of organisms and that transmits these properties from makeup of organisms. For example, geneticists and plant breeders often refer to the seeds and plants used in their research and breeding as their "collection of germplasm."

GH3 genes : In soybean and Arabidopsis, family of primary response genes that are stimulated by auxin within 5 minutes of treatment.

Gibberellic acid : Identical to gibberellin GA_3.

Gibberellin A_1 (GA_1) : A chemically distinct form of gibberellin. The primary active GA stem growth for most species.

Gibberellin A_3 (GA_3) : The main gibberellin found in fungal cultures; commonly available and used in the management of fruit crops, the malting of barley and the extension of sugarcane (to increase sugar yield). Occurs rarely in plants.

Gibberellin glycosides : Inactive or stored forms of gibberellin in which the hormone is covalently linked to a sugar (usually glucose).

Gibberellins :

- Class of phytohormones; the first gibberellin was found in the fungus Gibberella fujikora, from which these phytohormones derive their name; gibberellins are involved, e.g., in the promotion of seed germination, stem extension, and bolting.
- A large group of chemically related plant hormones synthesized by a branch of the terpenoid pathway and associated with the promotion of stem growth (especially in dwarf and rosette plants), seed germination, and many other functions.
- A class of plant growth substances first isolated from the fungus *Gibberella* fujikuroi found the rice.
- Possess gibbane skeleton consists of four rings designated as A, B, C, D required for growth of root, stem, leaves, coleoptile dormancy breaking, flower initiation, apical dominance, cell division, parthenocarpic fruit development, inducement of male sterility.

Girdling : Removal of a ring of bark from a woody stem that severs the vascular system.

Give rise to the suspensor.

Glabrous : Hairlessness as opposed to hirsute or hairy.

Glass : Solidlike liquid with an extremely high viscosity; examples of a glass are macaroni and "glass" as we know it from everyday life (which is not a solid, but a fluid, as apparent from the gradually changing properties of glass when it gets old); glass formation, rather than the formation of ice crystals, is essential to prevent

Glaucousness : Shiny appearance (of leaves), due to the presence of specific wax compounds.

Globular stage embryo :

- The first stage of embryogenesis. A radially symmetrical, but not developmentally uniform, sphere of cells produced by initially synchronous cell divisions of the zygote.
- Generation of an eight cell (Octant) globular embryo by 30 hrs after fertilization from zygote when apical cell undergoes a series of highly ordered divisions.

Gluconeogenesis :

- Conversion of storage lipids into sucrose i.e. lipolysis, â oxidation and glyoxylate cycle occur in the glyoxysomes.
- The biosynthsis of new carbohydrates from noncarbohydrate precursors.
- The synthesis of carbohydrates through the reversal of glycolysis.
- Synthesis of sugar from organic acids through reverse of glycolysis by certain organisms.

Glucoside (or glycoside) : Compound in which a side chain is attached to glucose by an acetal bond.

Glucosinolate : Secondary sulfur-containing metabolite in Brassicaceae (cabbage family) which gives these plants a distinct sharp smell and taste.

Glucosinolates (mustard oil glycosides) :

- Plant glycosides that breakdown to release volatile substances that defend against herbivores and are responsible for the oduor and flavour of broccoli, cabbage, and other cruciferous vegetables.
- Second class of plant glycosides called the glucosinolates or mustard oil glycosides, break down to release volatile defensive substances found principally in the Brassicaceae and related families, compounds responsible for the smell and taste of vegetables such as cabbage, broccoli and radish.

Glumes : Chafflike bracts; specifically, two empty chaffy bracts at the base of spike lets in grasses.

Glutamate dehydrogenase (GDH) : Catalyzes a reversible reaction that synthesizes a reversible reaction that synthesizes or deaminates glutamate as part of the nitrogen assimilation process.

Glutamate synthase :

- Enzyme that transfers the amide group of glutamine to 2-oxoglutarate. Yielding two molecules of glutamate. Also known as glutamine 2-oxoglutarate aminotransferase (GOGAT).
- Enzyme that converts ammonium into the amino acid glutamine.

Glutathione :

- The form in which assimilated sulphur is exported from leaves to sites of protein synthesis (shoot and root apices, and fruits). Signal molecule that coordinates root absorption of sulphate and shoot assimilation of sulphate.
- Tripeptide (g-glutamyl-cysteinylglycine) that acts as a reducing agent, protecting the cell against oxidative stress, and guards against chemical toxicity, via modification of (modified) xenobiotics.

Glyceroglycolipids : Glycerolipids in which sugars form the polar head group. Glyceroglycolipids are the most abundant glycerolipid in chloroplast membranes.

Glycerolipids :

- Lipids in which sugars form head group.
- Polar lipids that form the lipid bilayer of cellular membranes.

Glycerophospholipids (phospholipids) :

- Glycerolipids in which the polar head group contains phosphate. Most important glycerolipids in nonphotosynthetic tissue.
- Lipids in which the head group contains phosphate.

Glycine oxidation : The part of the photorespiratory carbon cycle in which glycine is converted into serine in the mitochondrial matrix, producing NADH, CO_2 and NH_4^+.

Glycolipid : Membrane lipid molecule with a short carbohydrate chain attached to a hydrophobic tail.

Glycolipide : A lipid containing a carbohydrate group.

Glycolipids : Containing one or more molecules of sugar.

Glycolysis or Embden - Meyer hof - parnas (EMP) Pathway :

- Oxidation of one glucose molecule forms two molecules of pyruvate and eight ATP molecules occurs in cytosol (cytoplasm without cell organalles).
- A series of reactions in which glucose is partly oxidized to produce two molecules of pyruvate. A small amount of ATP and NADH is produced.
- Ubiquitous metabolic pathway in the cytosol in which sugars are metabolized to pyruvate and/or malate with production of ATP and NADH (when pyruvate is the end product).

Glycophyte :

- Species restricted to nonsaline soils.
- Plants that are not able to resist salts to the same degree as halophytes. Show growth inhibition, leaf discolouration, and loss of dry weight at soil salt concentrations above a threshold. Contrast with halophytes.
- Plants which are sensitive to higher concentration of salts. Examples - Pulses.

Glycoprotein :

- A protein containing atleast one carbohydrate group.
- Any protein with one or more covalently linked oligosaccharide chains.

Glycoside (or glucoside) :

- Compound in which a side chain is attached to a sugar by an acetal bond.
- Compound containing an attached sugar or sugars.
- In which a sugar molecule is attached to the 3, 7 or 9 nitrogen of the purine ring, or to the oxygen of the zeatin or dihydrozeatin chain.

- Water soluble substances found in cell sap, on decomposition they yield glucose or some other sugar and some aromatic substances. Examples - Amygladin found in seeds of bitter almonds.

Glyoxylate cycle :

- Occurs in fat rich seeds containing glyoxysomes. Isocitric acid obtained from Kreb's cycle is directly oxidised into glyoxylate with formation of succinate which again enters in to Kreb's cycle.

- The sequence of reactions that convert two molecules of acetyl-CoA to succinate in the glyoxysome.

Glyoxysome :

- An organelle found in the oil rich storage tissues of seeds in which fatty acids are oxidized. A type of microbody.

- Microbodies found in fat rich seeds containing enzymes of glyoxylate cycle.

- Present in oil storing seeds contain glyoxylate cycle enzymes which help convert stored fatty acids into sugars.

GNOM gene :

- Arabidopsis gene for the development of roots and cotyledons. Homozygous GNOM mutant produces seedlings lacking both roots and cotyledons.

- Gene expression required for the establishment of axil polarity.

Golden rice : Transgenic rice seed with yellow pigment gene derived from chrysanthemum. The kernel appears yellow thereby deriving its name.

Golden yellow tag : The ISI certified colour of breeder seed tag (ISI No. 5-1978).

Goldman diffusion potential : The diffusion potential calculated from the Goldman equation.

Goldman equation : An equation that predicts the diffusion potential across a membrane, as a function of the concentrations and permeabilities of all ions (e.g., K^+, Na^+ and Cl^-) that permeate the membrane.

Golgi apparatus or dictyosome : Structure consisting of one or more stacks of three to ten flattened membrane sacs, or cisternae and irregular network of tubules and vesicles. Individual stack is called golgibody or dictyosome.

Gossypol : An aromatic 30-carbon sesquiterpene dimer that defends against insects, fungi, and bacterial pathogens in cotton.

Grain : Seed or seedlike fruit of any cereal grass, such as millet, oats, rice, and wheat.

Gram molecular weight : Molecular weight expressed in grams.

Grana lamellae : Stacked thylakoid membranes within the chloroplast. Each stack is called a granum, while the exposed membranes in which stacking is absent are known as stroma lamellae.

Grana /Granum :

- Stacked region of photosynthetic membranes (thylakoids) in chloroplasts that contains photosystem II with its light-harvesting complex.
- Lamellae form disc type of structures and arranged one above the other like stack of coins referred as grana lamellae. Individual is called granum and whole complex as grana.
- Stack of thylakoids (plural grana).

Granum thylakoid : (Thylakoid - Greek word means Thylacos - Sac or pouch)- Ends or margins of paired lamellae joined togather to form closed disc shaped sacs.

Gravitational water : Water percolates downward through larger pores between soil particles under gravity influence till it reaches water table. Not available for plant growth.

Gravitropic response : The signal that directs the growth of roots downward.

Gravitropism :

- Growth in response to gravity, enables roots to grow downward into the soil and shoots to grow upward away from the soil.
- Plant growth in response to gravity, enabling roots to grow downward into the soil and shoots to grow upward.

Gravity potential, Ψg : A component of water potential determined by the effect of gravity on the free energy of water.

Green fluorescent protien (GFP) : Visible marker that fuses with the phytochrome and facilitates the movement of phytochrome in the nucleus, that can be activated by light of an appropriate wave length being shown on plant cells.

Green house effect : The warming of Earth's climate, caused by the trapping of long wavelength radiation by CO_2 and other gases in the atmosphere. Term derived from the heating of a greenhouse resulting from the penetration of long wavelength radiation through the glass roof, the conversion of the long wave radition to heat, and the blocking of the heat by the glass roof.

Gross photosynthesis : Amount of carbon dioxide assimilated in chloroplasts; it is measured as net photosynthesis plus dark respiration.

Ground meristem :

- In the plant meristems, cells that will give rise to the cortical and pith tissues and in the root and hypocotyl, will produce the endodermis.
- Meristem that produces future cortex and endodermis.

Ground state or singlet state : Unexcited state of chlorophyll molecule before absorption of light.

Ground tissues : Includes parenchyma, collenchyma and scleranchyma.

Growth :

- Increment in mass, volume, length, or area of plants or parts thereof.
- Irreversible increase in size and volume of plant accompained by increase in dry weight.

Growth Analysis : Method of estimating net photosynthetic production.

Growth regulators : Some artificially synthesized compounds also perform same action as that of hormones. They along with hormones are called regulators. Examples - Indole butyric acid

(IBA), Naphthlene acetic acid (NAA), 2,4- Dichlorophenoxy acetic acid (2, 4-D), Para chloro phenoxy acetic acid (PCPA).

Growth respiration : Amount of respiration required per unit increment in biomass; it is not a rate.

Growth retardants : Inhibit gibberellin production which leads to dwarfness.

Growth velocity profile : The velocities of different tissue elements plotted against their distance from the apex to give the spatial pattern of growth velocity.

Growth zones : Regions with expanding tissues.

GSA gene : Gene of unicellular alga Chlamydomonas reinhardtii whose expression is mediated solely by a blue light sensing system. Encodes the enzyme glutamate-1-semialdehyde amino-transferase, a key enzyme in the chlorophyll biosynthesis pathway.

Guard cells :

- A pair of specialized cells which surrounds the stomatal pore associated with biological control.
- Specialized epidermal cells that surround the stomata and regulate the size of the stomatal pore.
- A pair of specialized epidermal cells which surround the stomatal pore and regulate its opening and closing.

Guttation :

- Oozing of water drops from the tip of leaves where principal vein ends, takes place from more or less spherical structures called hydathode or water stomata. Phenomenon is seen in oat, tomato, potato, garden nasturtium and grasses.
- An exudation of liquid from the leaves due to root pressure.
- Water exuded by leaves due to root Pressure.

Gymnosperm : An early type of seed plant. Distinguished from angiosperm by having seeds borne unprotected in cones.

❑❑❑

H

H⁺-PPase (proton pumping pyrophosphatase) : An electrogenic pump that moves protons into the lumen of the vacuole and the Golgi cisternae, energized by the hydrolysis of pyrophosphate.

Half life : Time taken to half disintegration.

Halogenation : Process of treating seeds with halogens (iodine, bromine) fumes or coating with halogens (chlorine or iodine) impregnated formulations to control seed deterioration caused by free –radicals. This process also offers protection against pests and pathogens to certain extent.

Halophyte :

- Species that typically grows on saline Soils.
- Able to grow in high salt concentrations.
- Plants that are native to saline soils and complete their life cycles in that environment. Contrast with glycophytes.

Hanging droplet method : Small drops (40 to 100 μL) of protoplast suspension placed on the inner side of the lid of a petridish. When the lid is applied to the dish, the culture drops are hanging or suspended from the lid.

Haploid : Chromosome no. (1 n)

Hard seeds :

- Seed with seedcoat impervious to water or oxygen required for germination, or to carbon dioxide that may accumulate within the seed and retard or prevent germination; sometimes overcome by scratching or scarifying the seedcoat or removal by brief immersion in concentrated sulfuric acid and thorough washing.
- Seeds that remain hard and firm at the end of the germination test due to impermeable seed coat. **Syn.** Impermeable seeds.

Hardening : Plant becomes resistant when exposed to stress conditions.

Hardiness - Innate ability of the plants to withstand or avoid injury from stress.

Harrington's thumb rule : This rule indicates the interaction of storage temperature and seed moisture content on the life span of stored seed. "For every 1% decrease in seed moisture the separate of seed is doubled and for every 10^0F (5^0C) drop in temperature the storage life of seed is doubled; this rule is applicable between 5-14% moisture content and 50-100^0F (0-50^0C)". **Syn.** Rule of thumb.

Hartig : Net hyphal network of ectomycorrhizal fungi that have penetrated intercellularly into the cortex of a higher plant.

Hartig net : A fungal network of hyphae that surround but do not penetrate the cortical cells of roots.

Harvest index (HI) : Ratio of economic yield to the biological yield (Synder and Carlson, 1984).

$$HI=\frac{\text{Economic yield}}{\text{Biological yield}}\times 100$$

Economic yield refers to the yield of economic valuable part of plant and biological yield total dry weight of plant.

Haustorium : Organ that functions in attachment, penetration, and transfer of water and solutes from a host to a parasitic plant.

Heart rot of Sugarbeet : Rotting of central portion due to Boron deficiency results in blackening and drying of central portion which extends towards out region.

Heart stage embryo :

- The second stage of embryogenesis. A bilaterally symmetrical structure produced by rapid cell divisions in two regions on either side of the future shoot apex.
- The stage whch forms through rapid cell divisions in two regions on either side of the future shoot apex. These two regions produce outgrowths that later will give rise to the cotyledons and give the embryo bilateral symmetry.

Heartwood : Central mass of xylem in tree trunks not functioning in water transport; it often contains substances that prevent decay and has a darker color than the surrounding sapwood.

Heat injury : Injury to seeds on exposure to heat during seed maturation or while drying at excessive temperatures.

Heat shock cognate proteins :

- Molecular chaperones contained in cells that are constitutively expressed and function like HSPs.
- Molecular chaperon proteins that are constitutively expressed and function like HSPs.

Heat shock proteins (HSPs) :

- A specific set of proteins that are induced by a repid rise in temperature, and by other factors that lead to protein denaturation. Most act as molecular chaperones.
- Production of a unique set of proteins in response to sudden, 5 to 10°c rises in temperature.
- Protein produced upon heat or other stresses.

Heat shock response : The increased synthesis of HSPs and the reduced synthesis of other proteins following a stressful but nonlethal heat episode.

Heavy metal : Metal with a mass density exceeding 5 g mL^{-1}.

Heliotropism, solar tracking :

- Movement of a leaf or flower that follows the angle of incident radiation.
- Movements of leaves toward or away from the sun.

Heme : Cyclic organic molecule that contains an iron atom in the center which binds O_2 in leghemoglobin and carries an electron in cytochromes.

Hemi celluloses :

- The mannans viz. Xylans, galactans etc. are hemicelluloses.
- Flexible polysaccharides that characteristically bind the surface of cellulose.
- Heterogeneous mixture of neutral and acidic polysaccharides, which consist predominantly of galacturonic acid and some rhamnose; these cell-wall polymers coat the surface of cellulose microfibrils and run parallel to them.
- Hetrogineous groups of cellulose like xylan, mannans, galactons.
- Heterogeneous group of polysaccharides that bind to the surface of cellulose, linking cellulose microfibrils together into a network. Typically solubilized by strong alkali solutions.

Herb : A plant with no enduring above ground parts, as distinct from trees and shrubs.

Herbicide bioassay test : A lab procedure that checks for the presence/ absence of herbicide resistance gene/ trait using viable seeds.

Herbivores : Plant feeding animals, including many species of insects and mammals.

Hermaphroditic flowers : Containing both carpels and stamens.

Hermetic storage : Storage in an airtight container.

Hermetically sealed :

- Sealed so that no gas or moisture can enter or escape.
- Sealed in such a way that moisture or gas can neither enter or escape.

Heterocyclic ring : A ring structure that contains both carbon and non-carbon (nitrogen or oxygen) atoms.

Heterocysts : Specialized, thick walled structures formed by cyanobacteria within plant cells to create an anaerobic environment for nitrogen fixation.

Heterodimer : Protein complex composed of two different polypeptide chains.

Heterokaryon or heterokaryocyte : Protoplast fusion from diverse sources to generate a novel cellular entity in which protoplasm, organelles and genetic elements from both the partners are put togather.

Heterotrimeric G protein : A membrane bound GTP binding protein composed of three subunits, á, â, and õ, mediates the signal transduction pathways of G-protein linked membrane receptors.

Heterotrophic : An organism or plant dependent of the external environment for food.

Heterotrophic cells : Organisms that rely upon reduced forms of carbon for the synthesis of structural elements and the supply of metabolic energy.

Heterotrophic growth : Growth of plants or parts thereof which depend on carbon supplied by another organism or organ of the plant.

Heterotrophic respiration : Respiration by nonautotrophic Organisms.

Hexokinase : Enzyme catalyzing the phosphorylation of hexose sugars while hydrolyzing ATP; a specific hexokinase is involved in sugar sensing.

High density polyethylene bag : Woven bags using HDPE strips that can be used for short - term storage of seeds.

Hilium :

- Scar on the seed coat marking the point of attachment to the ovary in angiosperms or to the megasporophyll in gymnosperms.
- Scar or point of attachment of the seed.
- Seed scar where the funiculus (the stalk of the ovule) was once attached.

HIR (high irradiance response) : Phytochrome responses whose magnitude is proportional to the irradiance (rather than fluence).HIRs saturate at fluences 100 times higher than LFRs and are not photoreversible. HIRs do not obey the law of reciprocity.

Histidine phosphotransferases : Enzymes which mediate the cytokinin signaling casade.

Histogenic layers : The tissue producing regions of the shoot apical meristem : tunica layers and corpus.

Historical filter : Historical factors that prevent a species from arriving at a site.

Hoagland solution :

- A nutrient solution for plant growth, originally formulated by Dennis R. Hoagland.
- Named affer D.R. Hoagland, a researcher who was prominent in the development of modern mineral nutrition research in the united states.

HOBBIT (HBT) gene : Expression of this gene may be an early marker of root meristerm identity (Willemsen *et al.* 1998).

Holoprotein :

- An apoprotein bound to a smaller, nonprotein molecule such as a chromophore.
- Combination of apoprotein and its chromatophore.

Homeobox : A sequence coding for the homeodomain, a 60-amino acid domain in transcription factors that binds to specific regions of DNA.

Homeobox genes : Genes encode homeodomain proteins that act as a transcription factors. These proteins play a major role in regulating developmental pathways in all eukaryotes.

Homeostasis : Tendency to maintain constant internal conditions in the face of a varying external environment.

Homeotic genes : First discovered in *Drosophila,* homeotic genes regulate the locations of body parts in flies. These transcription factors are typically characterized by homeodomains.

Homeotic mutations : In Drosophila, homeotic mutations in homeobox genes cause body segments and other structures to form at inappropriate locations. In plants, mutations with similar phenotypic effect on flower development, but do not involve homeobox genes.

Homodimer : Protein complex composed of two identical polypeptide chains.

Hormones :

- Organic compound produced in one part of a plant and transported to another, where it acts in low concentrations to control processes (phytohormone).
- Organic molecule often (but not always) synthesized in one location of the organism and transported to another, where it produces dramatic effects on growth or development at vanishingly low concentrations.
- Organic substances produced by the plants which in minute quantities increase, decrease or modify growth and development of plant. Site of synthesis is different from site of action . Examples - auxins, gibberellins, cytokinins, ethylene, abscisic acid.

Hot spots :

- Areas in a mass of stored seeds or grain where temperature is as high as 66° C or higher.
- Areas in stored seeds where temperature is 66^0C or higher caused by metabolic heating due to high moisture of stored seeds.

Hot water scarification : A physical scarification method of hard seeds by soaking in boiling water overnight to soften the thick seed coat. **e.g.** Prosopis, Luecaena.

Hourglass hypothesis : One hypothesis of how plants measure night length. Proposes that time is measured by a unidirectional series of biochemical reactions that start at the beginning of the dark period.

hph gene : Gene encoding hygromycin B resistance.

Hpt protein : In Arabidopsis, histidine phosphotransfer protein.

HSF : A specific transcription factor that acts during the transcription of heat shock proteins.

HSP 70 : Heat shock gene.

Hull : A tightly adhering outer covering of the seed or caryopsis.

Humic substances : High-molecular-weight polymers with abundant phenolic rings and variable side chains found in humus.

Humus : Amorphous soil organic matter.

Hundred seed weight : Weight of one hundred seeds, expressed in grams.

Husk : The outside hard envelope of fruit, which is rough and coarse as in caryopsis also for leaf sheath enclosing corn cob.

Huskless seeds : Seeds of paddy, Sunflower devoid of pericarp (husk); classified as pure seeds in physical purity test but restricted to a maximum permissible level of 2%.

H-value : Heterogeneity value.

hy mutants : Arabidopsis mutants in which white light does not inhibit hypocotyl growth as it does in wild type. Some hy mutants cannot synthesize phytochrome.

Hybridoma technology : Production of pure and specific antibodies in sufficient quantities.

Hydathodes : Specialized pores associated with vein endings at the leaf margin from which xylem sap may exude when there is positive hydrostatic pressure in the xylem. Also a site of auxin synthesis in immature leaves of Arabidopsis.

Hydration capacity : It is the difference in seed weight (g) of soaked and dry seeds.

Hydration index : It is the ratio of hydration capacity (100 seeds) to the dry seed weight (100 seeds) in grams.

Hydration-dehydration method : A technique developed by Prof. R.N. Basu to prolong the viability of stored seeds; seeds are soaked in water for specific period and dried back to original moisture content; this quenches the free radicals and reduces their damage to biomembranes.

Hydraulic : Lift upward movement of water from deep moist soils to dry surface soils through roots along a water potential gradient.

Hydraulic conductivity : Describes how readily water can move across a membrane and has units of volume of water per unit area of membrane per unit time, per unit driving force (i.e. m^3 m^2 S^{-1} MPa^{-1}).

Hydraulic conductivity of cell membrane : Describes how readily water can move across a membrane and has units of volume of water per unit area of membrane per unit driving force.

Hydrenchyma : Water-storing tissue; during dehydration of a plant, water is predominantly lost from the cells in the hydrenchyma, while other cells lose relatively less water.

Hydroactive closure : Stomatal closure in response to dehydration of whole leaf

Hydroactive stomatal closure : Closing of stomata in response to a closing signal which includes water stress. Depends on metabolic processes that reduce guard cell solute content resulting in water loss. Reversal of the mechanism of stomatal opening.

Hydrogen acceptors : Accepts hydrogen in the molecular form. Examples - NAD, NADP, FAD etc.

Hydrogen bonds :

- A weak chemical bond formed between a hydrogen atom and an oxygen or nitrogen atom.
- Weak electrostatic attraction between water molecules.

Hydrogen peroxide test : A quick test of viability for pine seeds; seeds are given a cut in the seed coat and soaked in 1% hydrogen peroxide and germinated; emergence of radicle is counted as germinable.

Hydrolysis : Cleavage of a covalent bond with accompanying addition of water, -H being added to one product and -OH to the other.

Hydrolyzable tannins : Tannins that are polymers of phenolic acids, especially gallic acid, and simple sugars. May be hydrolyzed by dilute acid.

Hydropassive closure :

- Closing of stomata due to loss of water directly from guard cells and the consequent loss of cell turgor. Operates at low external humidity, when direct water loss is not rapidly compensated.

- Stomatal closure due to decreased turgor as a result of direct loss of water by evaporation to the atmosphere.

Hydrophilic : Ability of an atom or a molecule to attract water molecules. For example, substances that can engage in hydrogen bonding are hydrophilic.

Hydrophilic (water loving) colloids : Affinity exists between particles of dispersed phase and dispersion medium. Example - Proteins, starch, gum, resigns, protoplasm, cellulose etc.

Hydrophobic (Water dreaders) :

- No affinity exists between particles of dispersed phase and dispersion medium. Examples - Rubber in water.
- Lacking a strong affinity for water.
- Substances, molecules, or functional groups that repel water molecules.
- The condition of certain compounds that lack affinity to water.

Hydrophyte :

- Plant that grows partly or wholly in water, whether rooted in the mud, as a lotus, or floating without anchorage, as the water hyacinth.
- Plants grow under water abudance.

Hydroponics :

- A technique for growing plants with their roots immersed in nutrient solution without soil.
- Technique of growing plants in nutrient solutions.

Hydropriming : Method of slow imbibition of seed by keeping them among moist substratum like cotton, paper, cloth at specified temperatures. Such seeds are dried back before sowing.

Hydroquinone (QH_2) : A fully reduced form of quinone.

Hydrostatic pressure : Pressure generated by compression of water into a confined space. Measured in units called pascals (Pa) or, more conveniently, megapascals (MPa).

Hydrotropism : Morphogenetic response (of roots) to a moisture gradient.

Hydroxyproline-rich proteins (HRGPs) : A class of wall structural proteins, rich in hydroxyproline, with possible roles in protection against pathogens and desiccation.

Hygrophyte : Species typically occurring on permanently moist sites.

Hygroscopic : Readily absorbing and retaining moisture, as caustic potash, or becoming coated with water, as glass.

Hygroscopic equilibrium measurement : Measuring the amount of water in or on a substance when there is no movement of water between the substance and the surrounding air.

Hygroscopic water : Water adsorbed on the surface of soil colloids in the form of tightly held thin film. Not available for plant growth.

Hyperpolarization : An increase in the normally negative inside, electrical potential (mV) that exists across the plasma membrane of a cell.

Hypersensitive response :

- A common plant defense following microbial infection, in which cells immediately surrounding the infection site die rapidly, depriving the pathogen of nutrients and preventing its spread.
- Rapid, localized cell death due to pathogen attack.

Hyphae (singular hypha) : The small, tubular filaments of fungi.

Hypocotyl :

- Part of seedling axis below cotylends.
- The region of the seedling stem below the cotyledons and above the root.

Hypocotyl hook : An inverted "J" formed when the apical end of the dicot hypocotyl bends back on itself as dicot seedlings emerge from the seed coat. Protects the shoot apex from damage during growth through the soil.

Hypogeal germination :

- Mode of germination in which the cotyledons (dicot) or scutellum (monocot) remains in the soil or within the seed. The shoot is carried above the soil level by elongation of epicotyl in dicots and mesocotyl in monocots. e.g. Most monocots; peas, Pigeon pea, Chickpea etc.

- Germination process in which cotyledons remain in the soil (Hypo - below, geal – earth)

Hyponasty : Upward bending of a plant organ.

Hypophysis : In seed plant embryogenesis, the apical-most progeny of the basal cell which contributes to the embryo and will form part of the root apical meristem.

Hypostomatous : With stomates at the abaxial (lower) side of the leaf only.

Hypoxia : Low oxygen concentration in (part of) a plant's environment.

Hypoxic : Refers to low concentrations (pressures) of oxygen. Contrast with anoxic.

❑❑❑

I

Ice nucleation :

- Process by which water molecules start to form a stable ice crystal. The surface properties of some polysaccharides and proteins facilitate ice crystal formation.

- Process whereby hundred of water molecules start to form a stable ice crystals.

Imbibition :

- Adsorption of water by hydrophilic colloids of plant material.

- The process of adsorption of water by seeds and hydration of seed tissue during the initial phase of germination. It is a physical process common to both live and dead seeds.

Imbibition damage : Damage caused due to uncontrolled imbibition by the 'gushing in' of water through micropyle / seed coat. This is found in most of the leguminous seeds causing damages to the tissues and affects germination.

Immature embryo : Condition in which morphologically immature embryo delays germination.

Immature seed : Seed that has not completely developed or reduced physiological and anatomical potential or reduced dry matter accumulation.

Immobilization : Nutrient absorption from the soil solution and sequestering by soil microorganisms.

Immunocytochemistry : The use of specific antibodies attached to identifying molecules in order to indicate the presence and perhaps location of molecules with antigenic properties.

Imperfect flowers : Flowers that lack either the male (stamens) or female (pistils) structures. Unisexual flowers.

Impermeability : The phenomenon noticed in hard seed coat that prevents the entry of air.

Impermeable seeds : Seeds that do not allow water and air to enter into them. **Syn.** Hard seeds.

Import : The movement of photosynthate in sieve elements into sink organs.

Imposed dormancy :

- Dormancy caused due to seed treatment injury or external factors. **Syn** Secondary dormancy, induced dormancy.
- Non germination of seeds due to absence of suitable conditions. Nikolaeva (1977).

In-vitro : Biological experiments performed in "the test tube" isolated from a whole organism. Contrast with in vivo.

In-vivo : Within the intact organism. Contrast with in vivo.

Incipient plasmolysis : Begining of pulling away of protoplast from the cell wall.

Incompatible : Interaction response of a resistant host to a virulent pathogen; interaction between pollen and pistil preventing sperm cells from reaching the ovule.

Incubation : Placing in controlled temperatures.

Incubation Drying Separation (IDS) method : A method usd to separate dead seeds from viable seeds based on the property that imbibe viable seeds loose moisture slowly upon drying as against dead seeds, which loose moisture rather fast. Hence after a definite period of drying following imbibition, if the seeds are placed in water, viable seeds being heavy sink to the bottom while dead seeds float and are separated.

Indeterminate growth :

- Activation of apical meristerm throughout growth period of the plant, continues to form new organs and tissues. Examples - Most of the pulses.
- Capacity for both vegetative growth and flowering over an extended period of time. Growth is not genetically limited, and will continue as long as environmental conditions and resources permit.

Indigo -carmine test : A viability test that stains only the dead cells, used similar to tetrazolium test.

Indirect embryogenesis : Development of embryo by cell proliferation i.e. callus from explant.

Indirect vigour tests : Those vigour tests that test certain seed qualities that are highly correlated with environment stress.

Indole : A precursor of IAA biosynthesis by a tryptophan independent pathway.

Indole-3-acetamide (IAM) : A biosynthetic intermediate for the synthesis of IAA in various pathogenic bacteria, such as Pseudomonas savastanoi and Agrobacterium tumefaciens.

Indole-3-acetic acid (IAA) : The most common naturally occurring auxin. When protonated, IAA is lipophilic and diffuses across lipid bilayer membranes, but when dissociated, the negatively charged IAA cannot cross membranes unaided. Free IAA is biologically active; IAA covalently bound to other molecules is inactive. A naturally occurring plant growth hormone.

Indole-3-acetonitrile (IAN) : An intermediate in one of three biosynthetic pathways for the synthesis of IAA in plants. IAN converted to IAA by action of nitrilase.

Indole-3-glycerol phosphate : A precursor of IAA biosynthesis by a tryptophan independent pathway.

Indole-3-pyruvic acid (IPA) : An intermediate in one of three biosynthetic pathways for the synthesis of IAA in plants. Formed by deamination of the amino acid tryptophan.

Indoxyl acetate test : A method to assess mechanical damage of soybean seed by soaking in 0.1% indoxyl acetate solution for two minutes followed by drying. Damaged seeds stain purpie.

Induced defense : Increased levels of plant secondary metabolites in response to herbivory or pathogen attack.

Induced dormancy : Imposed dormancy, Secondary dormancy.

Induced thermotolerance : Tolerance to lethal, high temperatures accomplished by periodic brief exposure to sublethal heat stress.

Inducible : The capacity for increased synthesis of a particular protein or proteins in response to a particular external singal, such as a hormone.

Induction period :

- Plant subjected for specific period of light , temperature and even chemicals .
- The period of time (time lag) elapsed between the perception of a signal and the activation of the response. In the calvin cycle, the time elapsed between the onset of illumination and the full activation of the cycle.

Inert matter : A component of purity analysis test; includes non-seed material and seed that do obey the definition of pure seed.

Infection thread :

- An internal tubular extension of the plasma membrane produced by the fusion of golgi dervied membrane vesicles at the site of infection.
- An internal tubular extension of the plasma membrane of root hairs through which rhizobia enter root cortical cells.

Infiltration : Movement of water into the soil.

Inflorescence meristem :

- Meristems produce bracts and floral meristems in the axils of bracts, instead of sepals, petals, stamens and ovules produced by floral meristems.
- Produces cauline leaves and inflorescence meristems in the axils of the leaves, as well as bracts and floral meristems in the axils of the bracts, and does not directly produce floral organs.

Infrared radiation : Radiation with wavelengths between approximately 740 nm and 1 mm; short-wave infrared is emitted by the sun (<3 mm), long-wave infrared is emitted at Earth temperatures (>3 mm).

Infra-red rays : Wavelengths longer than 760-100000 nm.

Infusion : The process of infusing the chemical into the seed using organic solvents.

Inner membrane : The inner of a mitochondrion's or chloroplast's two membranes.

Inoculum : Material such as spores; bacteria used as infecting agent with a disease or beneficial micro-organism.

Inositol : Intermediate products during conversion of one product to another. Generally found mixed with 9AA, Calcium, Magnesium, Potassium.

Intact seedlings : Well developed and healthy seedling with all essential structures.

Integral membrane proteins : Proteins embedded in, and often crossing, the lipid bilayer of biological membranes.

Integral proteins : Proteins embedded in the lipid bilayer in membrane.

Integument :

- The outer tissue layers surrounding the nucellus of an ovule; develops into the seed coat.
- The tissue covering the ovule, which upon maturation develops into seed coat.
- Thin sheaths over embryo sac.

Integumentary tapetum : A layer of nucellar cells with radial elongation and two nuclei that immediately surround and provide nutritive support to the developing embryo sac of some species. Later it hardens to form a protective layer around the seed.

Intercalary meristem :

- A meristem near the base of the internode that produces derivatives above and below.
- Meristem located near the base, rather than the tip, of a stem or leaf, as in grasses.
- Found within organs, aften near their basis.

Intercellular air space resistance : The resistance or hindrance that slows down the diffusion of CO_2 inside leaf, from the substomatal cavity to the walls of the mesophyll cells.

Intercepted solar energy (E_i) : Light intensities are converted into energy values.

Gasstra (1963) - 71 K Lux = 1 Cal. cm^{-2} Min^{-1}

$$\text{Hayashi (1966) Ei (\%)} = \frac{\text{Total incident energy - Transmitted energy}}{\text{Total incident energy}} \times 100$$

Interception : Acquisition of nutrients by roots as a result of growing through soil; the nutrients contained in the soil volume displaced by the growing root; precipitation water remaining in a plant canopy that does not reach the soil.

Intercrop : One crop plant grown in combination with at least one other crop on the same plot at the same time (e.g., an annual crop grown between trees).

Interference : Competition competition mediated by production of allelochemicals by a plant.

Intermediary cell :

- A type of companion cell with numerous plasmodesmatal connections to surrounding cells, particularly to the bundle sheath cells.
- Phloem cell in plants with a symplastic pathway of phloem loading; sucrose moves from the mesophyll into these cells, where

it is processed to form oligosaccharides that move to the sieve tube.

Intermediate day plant : A plant that flowers only between narrow day length limits (for example between 12 and 14 hours).

Intermediate filaments : Diverse group of helically wound fibrous elements, 10 nm in diameter, composed of linear polypeptide monomers of various types.

Intermembrane space : The fluidfilled space between the two mitochondrial or chloroplast membranes.

Internal conductance/resistance : Conductance/resistance for transport of CO_2 between the substomatal spaces and its carboxylation at the site of Rubisco in the chloroplast.

Internode : Portion of a stem between nodes.

Intussusception : Synthesis of cellulose and their inter liking.

Invertase : An enzyme that catalyzes the hydrolysis of sucrose to glucose and fructose.

Invigoration treatments : Seed treatment given to revitalize stored seeds by way of infusion of vitamins or growth regulators or antioxidants.

Ion channel/ion-selective channel : Pore in a membrane made by a protein, through which ions enter single file; channels are specific and either open or closed, depending on membrane potential or the presence of regulatory molecules.

Ionophore : A molecule that allows ions to cross lipid bilayers. There are two classes; carriers and channels. Carriers, like valinomycin, form cage like structures around specific ions, diffusing freely through the hydrophobic regions of the bilayer. Channels, like gramicidin, form continuous aqueous pores through the bilayer, allowing ions to diffuse through.

ipt 1 gene :

- The T-RNA gene involved in the cytokinin biosynthesis which encodes an isopentenyl transforase (IPT) enzyme that transfers the isopentenyl group from DMAPP to AMP (adenosine monophosphate) to form isopentenyl ribotide (Akiyoshi *et al.* 1984, Barry *et al.* 1984).

- The T-DNA gene involved in cytokinin biosynthesis; encodes cytokinin synthase, an isopentenyl transferase enzyme. Also called the tmr locus.

Irradiance :

- Amount of radient flux intercepted by a unit surface area. Radiometers are used to record.
- The amount of energy that falls on a flat sensor of known area per unit time. Expressed as watts per square meter ($W\ m^{-2}$). Note, time (seconds) is contained within the term watt : 1W = 1 joule (J) s^{-1}, or as moles of quanta per square meter per second ($mol\ m^{-2}\ s^{-1}$), also referred to as fluence rate.

Isobilateral : Having structurally similar upper and lower surfaces.

Isoflavonoids (Isoflavones) :

- Group of flavonoids in which position of one aromatic ring is shifted found mostly in legumes and have several different biological activities. Example - Retennoids which have insecticidal actions, others have anti estrogenic effects.
- Group of flavonoids with antimicrobial activity.

Isohydric : Maintaining a constant water status.

Isolines : Specially developed stocks to assess contribution of a physiological character to yield .

Isoprene (2-methyl-1,3-butadiene) :

- A gaseous, branched five-carbon molecule emitted by many plants. Isoprene emission protects leaves against damage from high temperatures. Provides the structure of the basic five carbon unit in terpene formation.
- Small unsaturated hydrocarbon, containing five carbon atoms (2-methyl-1,3-butadiene); volatile compound, synthesized from mevalonic acid and precursor of other isoprenoids; can be produced in large amounts by photosynthesizing tissue at high temperatures.

Isoprenoid lipids : Includes substances cholestrol, bile acids, vitamins, carotene and sex hormones.

Isotope : Atom of an element having same atomic number and chemical properties but differs in atomic weight.

Isotope discrimination : Alteration of the isotopic composition of an element via processes of diffusion, evaporation, and chemical transformation due to small differences in physical and chemical properties of isotopes; typically discrimination against the rare (heavy) isotope.

Isotope effect : End result of various processes that have different rate constants for different isotopes of the same element.

Isotope fractionation : Process that occurs when different isotopes of the same element have different rate constants for the same reaction or process, or chain of reactions or processes.

Isotropic :

- Characteristic of materials showing uniform structural and mechanical properties in all directions. Contrast with anisotropic.
- Similar in all directions.

Isozymes : Proteins that are structurally similar, but not identical, and share the same catalytic activity. Isozymes might be regulated by different mechanisms.

❏❏❏

Jasmonic acid

- Acid levels rise steeply in response to damage caused by a variety of different herbivores and trigger the formation of different kinds of plant defenses besides proteinase inhibitors including terpenes and alkaloids. It is known to induce the transcription of a host of genes involved in plant defense metabolism.
- A plant signaling molecule derived from linolenic acid (18 : 3) found in membrane lipids. Activates plant defenses against insects, fungal pathogens, and regulates plant growth including the development of anthers and pollen. Activates the expression of genes inovolved in response to various biotic and abiotic stresses, such as proteinase inhibitors.
- Secondary plant compound [3-oxo-2-(2′-cis-pentenyl)-cyclopropane-1-acetic acid], named after its scent from jasmine; stress signaling.

Juvenile phase :

- Stage in the life cycle of a plant between the seedling and reproductive phases; the vegetative phase in herbaceous plants; typically a period of rapid biomass accumulation.
- State of vigourous vegetative growth in plants and during this period they can not be induced to flower.

❑❑❑

K : Strategy suite of traits that enable a plant to persist in a climax community.

Kari-fisher titration method : A very accurate method of seed moisture estimation for pine seeds in which the moisture is extracted using methanol and titrated against a reagent.

Kcat (Catalytic constant of an enzyme) : Rate of the catalyzed reaction expressed in moles per mole catalytic sites of an enzyme (rather than per unit protein, as in Vmax).

Kernel : Edible part of cereals.

Ki : Concentration of an inhibitor that reduces the activity of an enzyme to half the rate of that in the absence of that inhibitor.

Kinases :

- Enzymes that have capacity to transfer phosphate groups from ATP to amino acids, such as serine or tyrosine, either on themseleves or on other proteins.

- Enzymes that have the capacity to transfer phosphate groups from ATP to other molecules.

Kinematics :

- The concepts and numerical methods applicable to the motion of fluid particles and the shape changes that the fluids undergo. Useful in analyzing meristematic growth.
- The study of the motion of fluid particles and the shape changes that the fluids undergo.

Kinetin :

- An adenine (or aminopurine) derivative, 6 furfurylaminopurine (Miller *et al.* 1955), which would stimulate tobacco pith parenchyma tissue to proliferate in culture.
- Substance originally isolated from autoclaved herring sperm DNA that greatly promotes cell division. Not a naturally occurring cytokinin. Chemically, a derivative of adenine (or aminopurine), 6 furfurylaminopurine.

Km : Substrate concentration at which a reaction proceeds at half the maximum rate.

Koh bleach test : A technique for identification of sorghum cultivars for dark tannin pigments by soaking seeds in a mixture of potassium hydroxide and liquid bleach in the ratio of 1 : 5.

Kranz anatomy :

- (G : kranz : wreath or halo). The wreathlike arrangement of mesophyll cells around a layer of large bundle sheath cells. The two concentric layers of photosynthetic tissue surrounds the vascular bundle. This anatomical feature is typical of leaves of C_4 plants.
- Specialized leaf anatomy of C4 species with photosynthetic bundle sheath cells surrounding vascular bundles.

Krebs cycle, tricarboxylic acid cycle, citric acid cycle :

- Metabolic pathway in the matrix of the mitochondrion oxidizing acetyl groups derived from imported substrates to CO_2 and H_2O.
- Oxidation of pyruvate yields CO_2, H_2O and 24 ATP molecules occurs in matrix between crystae in mitochondria.

❑❑❑

L

La mutant : Occurs in peas, when the cry^s mutation is present. Results in an ultra tall constitutive response mutant caused by the loss of function of a negative regulator.

Lactate dehydrogenase (LDH) : The enzyme that catalyzes the reversible conversion of pyruvate to lactate using the coenzyme NAD.

Lactic acid fermentation : Reaction in which pyruvate from glycolysis is reduced to lactate using NADH and thus regenerating NAD^+.

Lactophenol solution : A chemical concoction of latic acid : phenol : glycerin : water in the ratio of 1 : 1 : 2 : 1 used to prepare small grass seeds that cannot be bisected for tetrazolium staining; soaking in the solution makes the glume transparent revealing the stained embryo.

Lag phase of growth :

- Slow growth rate in the initial stages of plant development.

- During germination, the lag phase refers to the time after imbibition but before radicle protrusion when there is no water imbibition or extension of radicle.

Laminar air flow cabinet : Closed cabinet having bacterial HEPA (high efficiency perticulate air) filter, ultraviolet germicidal light etc. used for asceptic work in tissue culture.

Land race : Primitive crop varieties that are well adapted to specific conditions and form a chief source of genetic contamination in seed production.

Latent heat : Energy consumed or released by evaporation or condensation, respectively, of water (enthalpy of transformation); it results in respectively loss and gain of heat.

Latent heat of fusion : Energy required to melt 1 g of ice at O° C which is 335 Jules or 80 calories.

Latent heat of vaporization :

- Energy required to convert 1 g of water at 20° C to 1 g of water vapour at 20° C which is 2452 Jules or 586 calory.
- The energy needed to separate molecules from the liquid phase and move them into the gas phase at constant temperature.

Lateral bisection : A method of preparing small seeded grass for tetrazolium staining by moistening and cutting along the mid-axis of the embryo.

Lateral bud : Undeveloped shoot consisting of an axillary meristem, a short stem and immature leaves, often covered with bud scales and located above the point of attachment of a leaf to the stem.

Lateral meristems : Secondary meristems found in mature woody stems and roots as cylinders of meristematic cells. Their activity increases stem or root circumference.

Lateral roots (branch roots) : Arise from the pericycle in mature regions of the root through establishment of secondary meristems that grow out through the cortex and epidermis, establishing a new growth axis.

Latex : A complex, often milky solution exuded from cut surfaces of some plants that represents the cytoplasm of laticifers and may contain defensive substances.

Laticifers : In many plants, elongated, often interconnected phloem cells that contain rubber, latex and other secondary metabolites.

Law of limiting factor : When all the factors for photosynthesis are available in abundance the least available factor controls the photosynthetic rate, but if magnitude of this factor increases beyond saturation point other factors would become limiting (Blackman, 1905).

Law of minimum :

- When a physiological process is conditioned by several factors, the slowest factor controls the rate of process (Liebig, 1943).
- Obsolete concept that plant growth is always limited at any point in time by one single resource; it is not valid in this strict sense.

Law of reciprocity :

- The reciprocal relationship between fluence rate and the time.
- The reciprocal relationship between fluence rate ($molm^{-2}s^{-1}$) and duration of light exposure characteristic of many photochemical reactions as well as some developmental responses of plants to light. Total fluence depends on two factors : the fluence rate and the irradiation time. A brief light exposure can be effective with bright light; conversely, dim light requires a long exposure time.

Law of relatively limiting factor : According to Warburg, Boysen and Jensen (1918) it is not necessary that the factor which is quantitatively smallest will act as limiting factor, but the factor whose quantity is less than the actual amount required will act as limiting factor.

Le gene : Dominant allele for tall stems in peas, first studied by Mendel; codes for an enzyme, 3β- hydroxylase, that hydroxylates GA_{20} to produce GA_1.

Le mutant : In peas, a recessive mutant allele that causes a dwarf phenotype by interfering with the synthesis of active gibberellin. Mendel's dwarf gene.

LEA (Late Embryogenesis Abundant) Proteins :

- Proteins that play a role in cellular membrane protection.

- Proteins that stabilize cell membranes during desiccation. Encoded by a group of genes that are regulated by osmotic stress first characterized in desiccating embryos during seed maturation.

Leachates : Substances viz. sugars, free amino acids, organic acids and elements leached from seed with increase in permeability of seed membrance to ageing during imbibition of water by seeds.

Leaf area duration (LAD) : Active period of leaf growth or persistance of leaf area.

Watson (1952) $LAD = \frac{A_1 + A_2}{2} \times t_2 - t_1$ Area time

Leaf area index (LAI) :

- Total leaf area per unit area of Ground.
- Ratio of assimilatory surface area to the ground area.

Watson (1952) $LAI = \frac{A}{P}$

For two stages Gardner et al. 1985 $LAI = \frac{\frac{A1 + A2}{2}}{P}$

Leaf area ratio (LAR) :

- Ratio between total leaf area and total plant biomass.
- Ratio between leaf area and total plant dry weight Gregory (1926):

$LAR = \frac{A}{W}$ Area weight^{-1}

For two stages Gardner et al. 1985 $LAR = \frac{\frac{A_1}{W_1} + \frac{A_2}{W_2}}{2}$

Leaf boundary layer resistance : The resistance due to the layer of unstirred air next to the leaf surface through which water vapour must diffuse to reach the turbulent air of the atmosphere.

Leaf conductance/resistance : Conductance/resistance for transport of CO_2 or H_2O (vapor) of a leaf (it includes the conductance/ resistance for the stomatal and the boundary layer pathways in the case of H_2O, and additionally for the internal mesophyll pathway in the case of CO_2).

Leaf laminae : The blade of a leaf.

Leaf mass : Per unit leaf area (LMA) leaf mass expressed per unit leaf area.

Leaf-mass density : Leaf dry mass per unit of fresh mass or volume (in the first case, the presence of intercellular air spaces is not taken into account).

Leaf mass ratio (LMR), or leaf mass fraction (LMF) : Ratio of leaf and whole plant biomass.

Leaf primordial : Region of the shoot apical meristem that will form a leaf during the normal course of development.

Leaf stomatal resistance : The resistance associated with diffusion through stomatal pore.

Leaf turnover : Replacement of senescing leaves by new ones, not accounting for a change in leaf area.

Leaf weight ratio (LWR) - Ratio between leaf dry weight and plant dry weight

Watson (1952) $LWR = \frac{WL}{W}$

For two stages Gardner et al. (1985) : $LWR = \frac{\frac{WL_1}{W_1} + \frac{WL_2}{W_2}}{2}$

Leafy cotyledon1 (LEC 1) : A general repressor of vegetative development, expression is necessary throughout the embryogenesis (Lotan *et al.* 1998).

Lectins :

- Protein with noncatalytic sugar-binding domains; lectins are involved in defense and cellular interactions.
- Defensive plant proteins that bind to carbohydrates; or carbohydrate-containing proteins inhibiting their digestion by a herbivore.

Leghaemoglobin :

- An oxygen binding heme protein present in cytoplasm of infected nodule cells at high concentrations (700 μm in soybean nodules) and gives the nodules the pink colour. Its function is to transport oxygen to the respiring symbiotic bacterial cells (Ludwig and de Uries 1986).

- Pink pigment present in bacterial nodule acts as O_2 carrier to supply O_2 to Electron transport system, Kreb's cycle resulting in ATP synthesis.
- An oxygen-binding heme protein found in the cytoplasm of infected nodule cells that facilitates the diffusion of oxygen to the respiring symbiotic bacteria.
- Hemoglobin-like protein in nodules that associates with O_2 by means of a bound heme group.

Legume : A member of the family leguminosae often associated with rhizobia. Legumes include the pea (Pisum), clover (Trifolium), broad bean (Vicia), lentil (Lens), soybean (Glycine), kidney bean (Phaseolus), peanut (Arachis), and southern pea (Vigna).

Lemma :

- Refers to lower and outer of the two bracts enclosing the caryopsis in P*oaceae* family. The embryo in the seed lies on lemma side of the glumes.
- Small greenish bract that is part of the floret in grasses.

Lenticular transpiration : Loss of water from stem lenticells.

Leucoplasts :

- Lack pigments, starch, oil and protein containing are called amyloplasts, elaioplast and aleuroneplasts, respectively.
- Non pigmented plastids.

LFR (low fluence respose) : Phytochrome responses whose magnitude is proportional to low fluence (1.0 to 1000 ìmol m^{-2}). These include the classic red / far-red photoreversible responses.

LFY gene : In Arabidopsis, LEAFY (LFY) is a gene involved in establishing floral meristem identity.

LHCB gene family : Genes encoding the chlorophyll a/b-binding proteins (also called CAB proteins) of photosystem II. Regulated at the transcriptional level by both circadian rhythms and phytochrome.

Light :

- A form of radiant energy with properties of both particles and waves.

- Visible part of electromagnetic radiation occurs between the range of 390-760 mì or nm (1 m ì or nm = 10^{-9} meter or 10^{-7} cm).

Light channeling : In photosynthetic cells, the propagation of some of the incident light through the central vacuole of the palisade cells and through the air spaces between the cells.

Light compensation point :

- At certain light intensities the amount of CO_2 used in photosynthesis and CO_2 liberated in respiration remains equal.
- The amount of light reaching a photosynthesizing leaf at which photosynthetic CO_2 uptake exactly balances respiratory CO_2 release.
- Irradiance level at which the rate of CO_2 assimilation in photosynthesis is balanced by the rate of CO_2 production in respiration.

Light harvesting complex proteins (LHC proteins) :

- Chlorophyll-containing proteins associated with one or the other of the two photosystems in eukaryotic organisms. Also known as chlorophyll a/b antenna proteins.
- Complex of molecules of chlorophyll, accessory pigments, and proteins in the thylakoid membrane that absorbs quanta and transfers the excitation energy to the reaction center of one of the photosystems.

Light reaction :

- Takes place in grana results in production of ATP and $NADPH_2$ required for Co_2 reduction in dark reaction.
- Transfer of energy from absorbed light to ATP and NADP(H) in the photosynthetic membrane (thylakoid).

Light saturation (of photosynthesis) : Range of irradiances over which the rate of CO_2 assimilation is maximal and insensitive to level of irradiance.

Light scattering : The randomization of the direction of photon movement within plant tissues due to the reflecting and refracting of light from the many air-water interfaces. Greatly increases the probability of photon absorption within a leaf.

Light Screen : Device to show essentiality of light in the process of photosynthesis.

Light transmission ratio (LTR) : Ratio of light intensity within crop stand (I) and that incident upon plant (I_o).

Golinguai and Mabbayad (1969) $LTR = \frac{I}{I_0} \times 100$

Lignan : Phenolic compound with antifungal, antifeeding, and antitumor activity; minor component in most plants and tissues, but quantitatively more important in the wood of some tree species (e.g., redwood).

Lignin :

- A highly branched polymer of phenyl propanoid groups that plays both primary and secondary roles formed from three different phenyl - propanoid alcohals coniferyl, coumaryl and sinapyl alcohals which are synthesized from phenylalanine via various cinamic acid derivatives. It is found in the cell walls of various types of supporting and conducting tissues, notably the tracheids and vessel elements of xylem.

- Highly branched phenolic polymer with a complex structure made up of phenylpropanoid alcohols that may be associated with celluloses and proteins. Deposited in secondary walls, it adds strength allowing upward growth and Permitting conduction through the xylem under negative pressure. Lignin has significant defensive functions.

- Large amorphous polyphenolic polymer that confers woodiness to stems.

- The polysaccharide chains consist of alternating monosacchride units of N-acetyl-D-glucosamione and N- acetyl-nuramic acid a complex g-corbon sugar linked with 11 1-4 bonds.

Limiting factor : In any physiological process, the factor whose operation limits the rate at which the entire process can operate. For example, leaf photosynthesis can be limited by light or by ambient CO_2 concentrations.

Limonoids : Antiherbivore triterpenoids (30 carbons) which give citrus fruits their bitter taste.

Linear axis : In most plants, the body pattern in which the root and the shoot are at opposite ends.

Linoleic acid : Vitamin F, an unsaturated fatty acid containing 18 carbon atoms and three double bonds, that forms a major component of chloroplast cells.

Lipase : An enzyme that catalyses the hydrolysis of fats to glycerol and fatty acids.

Lipids :

- Esters of fatty acids and glycerol (alcohals).
- Group of substances comprising the fats and other esters that possess analogous properties.

Lipid bilayer :

- Description of biomembrane structure where two layers of hydrophobic fatty acid chains facing inside the membrane and hydrophilic phosphate heads lying outside the membrane.
- The core of cellular membranes formed by two layers of phospholipid molecules facing each other through their nonpolar tails.

Lipolysis : The process of hydrolysis of lipids and the breakdown of triglycerides to produce fatty acids that are used as metabolites.

Lipolytic degradation : Decomposition of fat.

Liquid droplet method : Suspending protoplasts in culture media and pipetting 100-200 μL droplets into 60 x 15 ml plastic petri dishes which are then sealed and incubated.

Liquid endosperm : Endosperm in liquid from. e.g. Coconut.

Liquid Nitrogen (LN_2) storage : Storing seeds in liquid nitrogen (-196^{0}C).

Liquid nitrogen (LN_2) vapour phase storage : A method of cryopreservation where packets are suspended in nitrogen vapours (-150^{0}C) instead of dipping in liquid nitrogen.

Liquid phase resistance : The resistance or hindrance that slows down the diffusion of CO_2 inside a leaf, from the walls of the mesophyll cells to the caroboxylation sites in the chloroplast.

Litter : Dead plant material that is sufficiently intact to be recognizable.

Litter quality : Chemical properties of litter that determine its susceptibility to decomposition, largely determined by concentrations of secondary metabolites and nutrients.

Little leaf disease : Leaves become distorted in shape and appearance due to Zn deficiency and clustered on short branches known as rosettes.

LN storage : Refers to storage of seeds in liquid nitrogen.

Lockhart equation : Equation that describes cell expansion in terms of turgor pressure and cell wall properties.

Lodging : The bending of cereal stalks to the ground because of the weight of moisture collecting on the ripened heads. Makes mechanical harvesting ineffective.

Log phase : The phase during germination during which there is increase in seedling elongation with increase in time.

Log phase of growth : Plant development stage in which growth increases very fast, also referred to as exponential growth phase or grand growth period.

Long day plants :

- Flowers when day length is more than certain critical light period (14 hours or more). Examples - Spinach, Sugarbeet, Letluce, Onion, Carrot, Barley, Wheat, Radish etc.
- A plant that flowers only in long days (qualitative LDP) or whose flowering is accelerated by long days (quantitative LDP).
- Plant whose flowering is induced by exposure to short nights.

Long - short day plants : Plants which require longer photoperiods for some days followed by short photo periods. Examples - Bryophyllum, Jasmine etc.

Long term storage : Implies storage of seed for 10-20 years or longer at 15% RH and 15^{0}C before being regenerated.

Long term storage of germplasm : As per criteria drawn by International Bureau of Plant Genetic Resources, seeds are expected to be stored for atleast 10-20 years before being multiplied again. Seeds are dried to a moisture content of 5% and stored in moisture proof glass jars with a sample size of 4000 for genetically uniform and 12000 for genetically un-uniform plants at - 18^{0}C.

Longevity : Long duration of life, length of life, or lifespan.

Longitudinal bisection : A method of preparing seeds for tetrazolium staining in case of linear seeds; seeds are moistened by soaking in warm water (30^{0}C) for four hours and then bisected along the long axis of the embryo and placed for staining.

Long-wave : Infrared radiation with wavelengths larger then approximately 3 mm emitted at Earth temperatures.

Looped structure : A classification of abnormal seedling where the hypocotyl coils around the seed.

Low constant temperature method : A method of determination of seed moisture where powered seeds are kept in hot air oven at 105± 2^{0}C for 17± 1hr.

Low moisture freezing limit : A term in cryopreservation denoting the lowest moisture content of seed for safe storage in liquid nitrogen.

Low temperature or chilling : Low temperature requirement for seeds (0 - 10° C) to release the dormancy while in a fully hydrated (imbibed).

Lumen cavity :

- Such as the space surrounded by the thylakoid membrane or the trap of Utricularia surrounded by cells.
- The cavity or within a tube or sac, especially the space inside thylakoid membranes.

Lux : An illuminance of one lumen per square meter.

LUX genes : Bacterial genes for luciferase that catalyze a light emitting reaction. Used as a reporter gene in transgenic plants in order to visibly indicate the activity of another gene sharing the same promoter.

Luxury consumption : Uptake of nutrients above the rate that enhances plant growth rate.

Lysigenous aerenchyma : Gas-transport tissue in plants that is formed from spatially selective death of expanded cells (see also schizogenous aerenchyma).

Lysimeter method : Gravimetric method to record transpiration rate.

Lysophosphatidylcholine : A product of PLA_2 phosphoilpid degradation that activates protein kinases; present on plant membranes in vitro and may be a second messenger in the auxin signal transduction pathway.

□□□

M

Macerase : Enzyme derived from fungus rhizopus used to dissolve pectin of middle lamella.

Macro or major elements : Elements required by plants in huge quantities i.e. 1000 mg/kg of dry matter.

Macrobiotic : A general term to describe the storage potential of seeds that can be stored safely for more than 15 years.

Macromolecule : Large molecule; grouping together of simple molecules capable of independent existence as in cellulose.

Macronutrients :

- Minerals obtained from the soil and present in plant tissues at concentrations usually greater than 30 ìmol g^{-1} dry matter. Nitrogen, potassium, calcium, magnesium, phosphorus, sulphur, silicon.

- Inorganic nutrients that a plant requires in relatively large quantities : K, Ca, Mg, N, S, P.

Macrosymbiont : Larger partner (i.e., higher plant) in a symbiosis with a microorganism.

MADS Box genes :

- Key regulators of important biological functions in plants, animals and fungi. Specific MADS box genes are important for developmental events in root, leaf, flower, ovule and fruit (Riehmann and Meyerowitz, 1977).
- Most plant homeotic genes belong to the class of related sequences.

Magnesium dechelatase : The enzyme that removes magnesium from chlorophyll as part of the chlorophyll breakdown process.

Maintenance respiration : Respiration required to maintain the status quo of plant tissues.

Malachite test : A viability test that uses the ability of dehydrogenase enzyme to decolourize malachite green upon soaking seeds in 0.004% malachite green solution for 24 hours.

Malate hypothesis : Also known as potassium ion exchange theory (Levitt, 1974). Opening and closing takes place in response to entry of K^+ ions in guard cells in day and removal of K^+ in night.

Male gametophyte : Body carrying male gamete.

Maleic hydrazide (MH) : (1,2 dihydro 3,6 pyridazinedione) A synthetic plant growth retardant that inhibits germination of seeds. It prevents the sprouting of onions, potatoes and prevents formation of suckers in tobacco. At lower doses it also acts as male sterilizing agent.

Malformation : Loss of lusture.

Malonaldehyde (MDA) : A product of lipid peroxidation of unsaturated fatty acid; used as an indicator or measure of peroxidation.

Malting : The first step in the brewing process. Germination of barley seeds at temperatures that maximize the production of hydrolytic enzymes.

Mass flow :

- Movement of ions, molecules or atoms in mass due to transpirational stream or pull.
- Movement of substances at equal rates as the fluid or gas in which they occur (e.g., transport of solutes in flowing water and CO_2 in flowing air).

Mass spectrometry (MS) : A method that identifies chemical compounds by their molecular mass to charge (m/z) rations and fragmentation patterns. MS can detect as little as 10^{-12} g (1 picogram, or pg) of IAA in plant extracts.

Mass transfer rate : The quantity of material passing through a given cross section of phloem or sieve elements per unit time.

Material : Pertaining to matter, the amount and kind of matter under consideration.

Matric potential, $Ø_m$ (or matric pressure) :

- The sum osmotic potential ($Ø_s$) + hydrostatic pressure ($Ø_p$). Useful in situations (dry soils, seeds, and cell walls) where the separate measurement of $Ø_s$ and $Ø_p$ is difficult or impossible.
- Measure of decrease in free energy due to affinity of water molecules to the colloidal substances of plant materrial, represented with (-) sign.
- Component of the water potential that is due to the interaction of water with capillaries in large molecules (e.g., clay particles in soil).

Matri-conditioning : A from of osmotic priming using absorbent solid material like calcinite clay or Microcel E (a patented product); the matrix is soaked in required chemicals and seeds are placed for priming. The ill effects of water are avoided in this technique. Syn. Solid-matrix priming.

Matrix :

- The compartment enclosed by inner membrane, contains the enzymes of Kreb's cycle.
- A substance in which other structures or organelles are embedded; used for the compartment inside chloroplasts or

mitochondria, not including the membrane system; also used for the substance in which cell-wall macromolecules are embedded.

- The colloidal aqueous phase contained within the inner membrane of a mitochondrion.

Maturation stage embryo : Towards the end of embryogenesis, the embryo and seed loose water and become metabolically quiescent as they enter dormancy.

Maturation zone : The region of the root that has completed its differentiation and shows root hairs for the absorption of water and solutes; and competent vascular tissue.

Maximum permissible limit : Refers to the maximum permitted contamination in seed field as per Seeds Act, 1966; fields showing contamination beyond this are summarily rejected.

MCP : A strong competitive inhibitor of ethylene binding 1-methylcyclopene.

Mean daily germination : A measure of seedling vigour; calculated as the cumulative percentage of full seed germination at the end of the test period and divided by the number of days from sowing to the end of the test.

Mean germination time : Time taken for 50% of the seeds to germinate. It is an indication of vigour.

$$MGT = \sum ni \times d/Ni$$

Where, ni=number of germinated seeds on each day

D=number of days from beginning of test

Ni=total number of germinating seeds.

Mean residence time (of a nutrient in a plant) : Time a nutrient remains in the plant, between uptake by the roots and loss (e.g., due to leaf shedding, consumption by a herbivore).

Mechanical admixture : A cause of varietal deterioration due to accidental admixture of varieties at field, during harvest, drying or processing.

Mechanical injury : Injuries caused to seed by faulty handling; improper machine setting that causes cracks on seed surface or damages the seed.

Mechanical scarification : A method of scarification done using mechanical means like making an incision or hole using blade or needle, scratching seed against a rough surface or clipping the seed coat using a nutcracker, nail cutter or knife.

Media preparation area : Area available for keeping chemicals, glassware, culture vessels, closures, refrigerator, pH meter, balance, hot plot, stirrers, gas source, oven, autoclave or pressure cooker, microwave etc.

Medium term storage : As per criteria defined by International Bureau of plant Genetic Resources, seeds are to be stored for 5-10 years before regeneration at 35% RH or lower at 0-10^{0}C.

Megagametogenesis : The process of formation of embryosac from megaspore.

Megapascals (MPa) : 10^6 pa

Megasporangium : Body containing megaspores.

Megaspore : Formed by the meiotic division of megaspore mother cell; although four are formed initially only one survives which develops into embryosac by megagametogenesis.

Megaspore mother cell : Body in the ovule associated with formation of egg cell.

Membrane : Any thin, soft, pliable sheet or layer of animal or vegetable origin; a limiting protoplasmic surface.

Membrane channel : Transmembrane protein complex that allows inorganic ions, small molecules, or water to move passively cross the lipid bilayer of a membrane.

Membrane (phospholipid) : Bilayer that surrounds cells (plasmalemma), cell organelles, and other cell compartments.

Membrane fluidity : Loose term to describe the extent of disorder and the molecular motion within a lipid bilayer; fluidity is the inverse of viscosity.

Membrane permeability : The extent to which a membrane permits or restricts the movement of a substance.

Mericarp : One halp of a schizocarp, which splits open upon maturation. It is considered as a pure seed in purity analysis of *Umbelliferae crops.*

Meristem :

- A group of cells capable of dividing indefinitely.
- Localized regions of cell division where plant growth is concentrated.

Meristem and shoot tip culture : Culture of a dome of tissues located at the extreme tip of a shoot and measures Ca 0.1 mm in diameter and Ca. 0.25 to 0.3 mm in length.

Meristem identity genes :

- Genes necessary for the initial induction of the floral organs.
- Genes necessary for the initial induction of the organ identity genes. Theses genes are the positive regulators of floral organ identity.

Meristem tip culture : Inoculation of shoot meristem of varying sizes on to agarized nutrient medium.

Meristematic cells : Cells under rapidly dividing state.

Meristematic zone : A region at the tip of the root containing the meristem that generates the body of the root. Located just above the root cap.

Mesobiotic : A general term to describe the storage potential of seeds that can be stored safely for 3-15 years.

Mesocarp : The middle layer of the fruit wall of berry or a drupe.

Mesocotyl :

- In coleoptile elongation of region between its base and the attachment of scutellum.
- Part of embryonic axis lying in between the shoot and root apices; it is a compound structure formed by adanation of part of the scutellum to the embryonic axis and is the orign of seminal roots.

Mesophyll cells :

- Cells between upper and lower epidermis except cells of vascular bundles.

- Photosynthetic cells found in the mesophyll of leaves of both C_4 and C_3 plants.
- Leaf tissue found between the upper and lower epidermal layers, consisting of palisade parenchyma and spongy mesophyll.
- Photosynthetic cells in a leaf; in a dorsiventral leaf often differentiated in palisade and spongy parenchyma cells.

Mesophyte :

- A plant surviving on moderate quantity of water.
- Plant that typically grows without severe moisture stresses.
- Grow where the amount of water is intermediate.

Messenger RNA (mRNA) : Form of ribonucleic acid which conveys the genetic messages coded in the DNA of the nucleus to the ribosomes of the endoplasmic reticulum where proteins are synthesized.

Metabolic absorption : Metabolic energy is required.

Metabolism : Chemical changes occurring in an organic body in living and growing processes.

Metabolites : Products of metabolism.

Metallophyte : Species that typically grows in areas with high concentrations of certain heavy metals in the soil.

Metallothionein : Low-molecular-mass metalbinding Protein.

Methanogenic bacteria : Obligate anaerobes that use CO_2 and H_2 as electron acceptor donor, respectively, in anaerobic respiration for the production of methane.

Methemoglobinemia : A disease of humans and livestock caused by the consumption of plant material high in nitrate. The liver reduces nitrate to nitrite, which combines with hemoglobin blocking its ability to bind oxygen.

Michaelis-Menten constant (K_m) : A constant in the Michaelis-Menten equation for enzyme kinetics. The constant reflects the binding affinity of the substrate for the enzyme or a solute for its carrier and corresponds to the concentration of substrate that gives ½ the maximum velocity that the enzyme can catalyze.

Micro tubers : A system of multiplying or culturing potato shoots in glass jars to produce miniature tubers, which are used for planting; these are virus free.

Micro, Minor or Trace elements : Elements which are required in less quantities i.e. 100 mg/kg of dry matler, Example Iron, Molybdenum, Copper, Chlorene, Boron, Manganese, Zinc and Nickel.

Microaerobic conditions : Reduced oxygen conditions maintained by some aerobic nitrogen-fixing bacteria such as Azotobacter through high rates of cellular respiration.

Microbiotic : A general term to describe the storage potential of those seeds whose life span does not exceed beyond three years.

Microbodies : A class of spherical organelles surrounded by a single membrane specialized for one metabolic function.

Microclimate : Local atmospheric zone where the climate differs from the surrounding atmosphere. (e.g., near a leaf, within a forest and near a body of water).

Microfibril : Structural component in cell walls, consisting of bundles of around 50 cellulose molecules, that provides the tensile strength of the wall.

Microfilament :

- A component of the cell cytoskeleton made of actin; it is involved in organelle motility within cells.
- Solid, with a diameter of 7 nm, composed of a special form of protein found in muscle, globular actin or G - actin.

Microgametogensis : The process of formation of microspore.

Micronutrient :

- Minerals obtained from the soil and present in plant tissues at concentrations usually less than 3 μmol g^{-1} dry matter. Chloride, iron, boron, manganese, sodium, zinc, copper, nickel, molybdenum.
- Inorganic nutrients that a plant requires in relatively small quantities : Mo, Cu, Zn, Fe, Mn, B, Cl (see macronutrients).

Microprojectile (particle gun) : High speed microprojectiles usually made of gold or tungsten coated with the DNA molecules to be introduced into the plant cells, through propelling it in to the plant cell by an electric discharge or cartridge blast.

Micropropagation : Clonal propagation in vitro.

Micropyle :

- A pore in the ovule wall of the integument through which the pollen tube enters to fertilize the egg; in the seed it the point of entry of free water during imbibition.
- Aperature in seed which absorbs water during imbibition.
- Pore or opening through which the pollen tube enters the embryo sac during fertilization and through which the radicle fre quently emerges when the seed germinates.

Microsporangium : Body containing microspores i e pollen grains.

Microspore : The male spore in the anther from which the male gametes develop.

Microspore mother cell :

- Cell in the anther lobe connected with production of pollen grains.
- One of the many cells in the microsporangium (anther), which undergoes microsporogensis to yield four microspores (pollen).

Microsymbiont : Smaller partner (i.e., microorganism) in a symbiosis with a higher plant.

Microtubule :

- Component of the cell cytoskeleton made of tubulin, a component of the mitotic spindle, and a player in the orientation of cellulose microfibrils in the cell wall.
- Hollow cylinders with an outer diameter of 25 mm composed of polymers of protein tubulin.

Mid storage correction : Those treatments given to stored seeds in the middle of their storage period (6-8 months) so as to improve their performance.

Middle lamella :

- A thick layer consists of calcium and magnesium pectate through which each walled cell and its neighbour are connected togather.
- A thin layer of pectin rich material at the junction where the primary walls of neighboring cells come into contact. Originates as the cell plate during cell division.

Milk stage : Stage in the development of the kernel of a grain when its contents are a milklike liquid.

Mimicry : Resemblance of an organism to another organism or object in the environment, evolved to deceive predators, prey, pollinators, etc.

Mineral nutrients : Inorganic ions absorbed from soil.

Mineralization :

- Breakdown of organic matter releasing inorganic nutrients in the process.
- Breaking of organic compounds by action of soil micro organisms to release nutrient elements.
- Process of breaking down organic compounds by soil microorganisms that releases mineral nutrients in forms that can be assimilated by plants.

Mistletoe : Xylem-tapping stem parasite.

Mitochondria :

- Spherical or tubular structure in cells, have a smooth outer membrane and a highly convoluted inner membrane, sites of respiratiion where ATPs are formed from ADP and inorganic phosphate by using energy released through normally carbohydrate metabolism.
- Organelle in which part of the respiratory process (tricarboxylic acid cycle, respiratory electron transport) occurs.

Mitosis : Process by which previously replicated chromosomes are aligned, seperated and distributed in an orderly fashion to daughter cells.

Modified halving method : A method of mixing and dividing seeds by pouring the seeds on cubical cups which alternatively lack bottom, the seeds retained in these cups with bottom are used as a sample for analysis.

Moiety : A part of a larger molecule or structure.

Moist storage : A method of storing recalcitrant seeds by keeping them in moist sawdust or charcoal to prolong the viability. **Syn.** Imbibed storage.

Moisture content equilibrium : State when the moisture content in the air and in the seed remains constant; no movement of moisture either into or out of the seed.

Moisture equilibrium :

- A situation where the moisture content of the surrounding air and that of the seed remain constant when further exchange ceases.
- Relative humidity of the air is in equilibrium with the moisture content of the seed.

Moisture gradient :

- Difference in moisture concentration inside and outside a seed and between the seed surface and the air surrounding the seed.
- The difference prevailing in the moisture content of the seed and that of the surrounding air; steeper the gradient higher will be the exchange rate.

Moisture pervious container : Seed packing material that allows free exchange of moisture between the storage atmosphere and the seed. **e.g.** Cloth, Gunny.

Moisture resistant container : Packing material that prevents moisture migration but allows moisture vapour between the container and seeds. **e.g.** Polythene bags of less than 600 gauge; HDPE bags.

Moisture translocation : Movement of moisture from one part of the seed to another in storage container.

Moisture vapour proof container : Packing material that prevents the exchange of moisture between seeds and the storage atmosphere. E.g. polythene bag > 700 guge thickness; rubber stopper glass bottles; paper aluminum foil laminated polythene pouces.

Molal solution : Gram molecular weight of a substance dissolved in 1 kg of solvent.

Molar solution : Gram molecular weight of a substance dissolved in a solvent to make final volume 1 litre.

Molecular motors : Large protein complexes associated with microtubules that move flagella, vesicles, chromosomes, and the cellulose synthase complex.

Molecular nitrogen : Found in earths crust, rocks and sediments (17.5 - 18.4 x 10^{15} tons) and in atmosphere (3.5 x 4.0 x 10^{15} tons). Certain bacteria and blue green algae also fix atmospheric nitrogen.

Monocarboxylic acid : Possess single carboxyl (COOH) group. Examples-glycolic acid, pyruvic acid, glyceric acid.

Monocarpic : Life cycle that ends after a single seed production event; the plant flowers only once during its lifetime, which can be after several years or even decades of vegetative growth.

Monocarpic senescence : Senescence of the entire plant after a single reproductive cycle, initiated by fruit and seed development. Can be prevented by continual removal of flowers.

Monocot :

- An abbreviated name for monocotyledons referring to the seeds having single cotyledon (scutellum). **e.g.** Rice, wheat etc.
- One of the two classes of flowering plants characterized by a single seed leaf (cotyledon) in the embryo. Contrast with dicot.

Monoecious : Refers to plants in which male and female flowers are found on the same individuals, e.g., cucumber (*Cucumis sativus*) and maize (*Zea mays*). Contrast with dioecious.

Monoglyceride : One hydroxyl group of glycerol combines with fatty acids.

Monooxygenases : Group of oxygenase enzymes that incorporate only one of the oxygen atoms from O_2 into a carbon compound; the other oxygen atom is reduced to water using NADH or NADPH as an electron donor.

Monopteros gene :

- Gene required for the formation of the embryonic primary root, but not for root formation in the adult plant, also required for the formation of vascular tissue in postembryonic development. (Przemeck *et al.* 1966).
- Arabidopis gene whose mutations produce seedlings lacking both a hypocotyl and a root, but have cotyledons and a shoot apical meristem.

Monosaccharides : Simple sugar, consists of a single polyhydroxy aldehyde or ketone unit.

Monoterpenes : Terpenes having ten carbons, two five-carbon isoprene units.

Morphogenesis : During development plant transforms itself from a fertilized egg to a plant. Originated from Greek word i.e. Morpho-form, genesis - origin.

Morphological maturity : An index used for certification of mature seed based on complete development of embryo as opposed to physiological maturity.

Mosaic growth : Synthesis of cell wall material at localized area, scattered over the surface. In this case cytoplasm pushes apart the microfibrils and the s ces which remain unfilled becomes pit filled.

Motor proteins : Proteins that drive cytoplasmic streaming and the movement of organelles within the cell.

Mottling of wheat seed : Caused due to deficiency of nitrogen leading to reduced weight, reduced water absorption and low market value.

Movement proteins : Viruses genome encode proteins that facilitates the movement of viral RNA genome through plasmodesmata.

MS (Murashige and Skoog (1962), medium : Salt composition widely used in different tissue culture systems.

Mt DNA : Mitochondrial DNA.

Mucilage :

- A slimy secretion on seeds, which swells in contact with water.
- Composed mainly of polyuromides and galactomides.

Multinate growth : Apposition of microfibrils with the early layers modyfied into the microfibrilar orientation during cell enlargement.

Mutualism : A symbiotic relationship in which both organisms benefit.

Mycelium : The mass of hyphae that forms the body of a fungus.

Mycorrhiza (plural is mycorrhizae or mycorrhizas) :

- Structure arising from a symbiotic association between a mycorrhizal fungus and the root of a higher plant (from the Greek words for fungus and root, respectively).
- A symbiotic association of the roots of some plants and fungi.
- Fungi associated with root system facilitates nutrient uptake by roots.
- (Sigular mycorrhiza, from the Greek words for "fungus" and "root"). The symbiotic (mututalistic) association of certain fungi and plant roots. Facilitate the uptake of mineral nutrients by roots.

Mycorrhizal dependency (of plant growth) : The ratio of dry mass of mycorrhizal plants to that of plants of the same genotype grown without mycorrhizal fungus under the same environmental conditions.

Mycotoxins : Toxic substances produced by fungi. **e.g.** Aflatoxin.

Myosins : Proteins that have the ability to hydrolyse ATP to ADP and Pi when activated by binding to an actin microfilament.

❑❑❑

N

N gene : Gene producing the resistance against tomato spotted with virus.

Na gene : Which conditions narrow leaves.

Na mutant : In peas, a recessive mutant allele causing an extreme dwarf phenotype by completely blocking gibberellin biosynthesis no GA_{12}- aldehyde is produced.

Naked seed : General term used for Gymnospermous seeds that are not covered by ovary wall.

Naked stratification : Prechilling of seeds without use of moisture holding medium.

Natural products : See secondary metabolites.

Necrosis : Death of tissues brought about by physical damage, poisions or other external injury.

Necrotic lesion :

- Formation of a small circular island of cell death from the dead cells.
- Regions within an organ in which cells die while surrounding regions remain alive.

Necrotic spots :

- Dark green colouration of the leaves, which may be malformed and contain small spots of dead tissue.
- Small spots of dead leaf tissue. For example, a characteristic of phosphorus deficiency.

Nectar : Sugary fluid produced in specialized glands of some flowers to attract insects as pollinators. Sometimes contains amino acids and other nitrogenous substance.

Nectar guides : On a flower, symmetric patterns of stripes, spots, or concentric circles formed by flavonols and other compounds to attract and orient pollinating insects.

Negative geotropism : Upward bending of roots away from the substratum either due to moisture deficit in the medium or due to medium phytotoxicity.

Nernst equation : An equation that predicts the electrical potential at which a charged ion will be in equilibrium across a membrane, as a function of the relative concentrations of the ion on each side.

Nernst potential : The electrical potential described by the Nernst equation.

Net assimilation rate (NAR) :

- Rate of plant biomass increment per unit leaf area; synonym is unit leaf rate (ULR).
- Rate of measurement of photosynthesis per unit leaf area per unit time.

$$\text{Gregory (1926) NAR} = \frac{W2 - W1}{A2 - A1} \times \frac{\text{Ln } A_2 - \text{Ln } A_1}{t_2 - t_1} \text{ Weight Area}^{-1}\text{Time}^{-1}$$

Where A_1 & A_2 area leaf area at two intervals.

Net ecosystem carbon balance (NECB) : Net change in ecosystem carbon content due to all processes, including photosynthesis, respiration, loss of biomass, leaching, and lateral movements and transfers.

Net ecosystem production (NEP) : Organic carbon accumulation that equals gross photosynthesis minus ecosystem respiration or net primary production minus heterotrophic respiration.

Net photosynthetic production : Amount of photosynthates left after the respiration and other biochemical reactions.

Net primary production (NPP) : Quantity of new plant material produced annually per unit ground area including lost plant parts; equals gross photosynthesis minus autotrophic respiration.

Net Reproductive Effect : Refers to the proportion of plants annual assimilated efforts devoted to reproduction.

$$\text{NRE} = \frac{\text{Dry weight of reproductive parts}}{\text{Total dry weight of the plant}}$$

Neuton : Unit of force in MKS (Meter, Kilo and Second) system.

Night break :

- An interruption of the dark period with a short exposure to light that makes the entire dark period ineffective.
- Influence of dark period is made ineffective by interruption with a short exposure to light. night with open stomata.

NIT gene : One of the genes (NIT 1 through NIT4) encoding nitrilase enzymes that convert IAN to IAA.

Nitrate nitrogen : The form of nitrogen i.e. NO_3 easily absorbed by roots.

Nitrate reductase : Enzyme that reduces nitrate (NO_3^-) to nitrite (NO_2^-). Catalyzes the first step by which nitrate absorbed by roots is assimilated into organic form.

Nitrification :

- Conversion of ammonia to nitrate form by bacteria of Nitrosoman and Nitrobacter group.

- Microbial process that transforms ammonia, via nitrite, into nitrate.

Nitrite reductase : The enzyme that reduces nitrite (NO_2^-) to ammonium (NH_4^+).

Nitrogen assimilation : Incorporation of inorganic nitrogen (nitrate, ammonium) into organic compounds.

Nitrogen containing compounds : Secondary metabolites possess nitrogen in their structure. Examples - Alkaloids, cyanogenic glycosides, glucosinolates and nonprotein amino acids.

Nitrogen fixation :

- Breaking of an exceptionally stable triple covalent bond between two nitrogen atoms (N = N) to produce ammonia (NH_3) or nitrate (NO_3^-).
- Reduction of dinitrogen gas to ammonium by specialized microorganisms.
- The natural or industrial processes by which atmospheric nitrogen N_2 is converted to ammonia (NH_3) or nitrate (NO_3^-).

Nitrogenase : The enzyme complex that catalyzes the reduction of N_2 to 2 NH_3. The reaction is coupled to the reduction of two protons to H_2.

Nobbe trier : A long pointed hollow tube with an oblong slit near the pointed end used for sampling seeds from bags; named after Dr. Fedrich Nobbe, the father of seed testing.

Nod factors :

- Product of nod genes required for successful nodulation in the legume–rhizobium symbiosis.
- Lipochitin oligosaccharide signal molecules, all of which have a Chitin B - 1 - 4 linked N-acetyl-d-glucosamine backbone and a fatty acyl chain on the C-2 position of non reducing sugar.

Nod gene : Rhizobial gene involved in the process of nodulation.

Nodal or brace roots : Adventitious roots originate from the nodes.

Node : Position on the stem where leaves are attached.

Nodulation : Formation of nodules in symbiotic N_2-fixing plants.

Nodulation (nod) genes :

- Rhizobial genes that participate in the nodule formation (Heidstra and Bisseling 1996).
- Rhizobial genes, the products of which participate in nodule formation.

Nodule primordium : Rhizobiainfected root cortical cells that dedifferentiate and start dividing.

Nodules :

- Tumour like structure on root formed due to bacterial (rhizobium) infection.
- The special organs of plant host that enclose the nitogen fixing bacteria.
- Specialized organs of a plant host containing symbiotic nitrogen-fixing prokaryotes.

Nodulin genes (Nod) : Plant genes specific to nodule formation.

Nodulins : Class of plant proteins that are synthesized in legumes upon infection by rhizobia.

Nomograph : A chart or diagram of a scaled lines or curves used to help in calculations. It has three or more scales in which one line joining values on line two, helps to determine the value of a third.

Non – free flowing seeds : Seeds that do not flow freely when dropped from a container. **e.g.** Grasses, fuzzy cotton seeds.

Non climacteric fruits : Do not show increase or decrease in respiration rate during storage. Examples - Cherry, Cucumber, Grape, Pineapple, strabery, sweet orange etc.

Non cyclic photophosphorylation : PSI gets back electron from PS II and PS II from photolysis of water. Hence, the movement of electrons is not in a cyclic fashion.

Non electrolytes : Substances which do not dissociate into ions when dissolved. Examples - Sugar etc.

Non protein amino acids : Many plants contain unusual amino acids that are not incorporated into proteins but are present instead in the free form and act as protective substances. Examples - Canavanine, azetidine - 2- carboxylic acid.

Nonclimacteric fruits : Fruits such as citrus fruits and grapes that do not exhibit the increased cellular respiration and ethylene production rise seen in climacteric fruits.

Non-inductive conditions : External conditions under which a plant continues to remain in vegetative phase.

Nonphotochemical quenching : The quenching of chlorophyll fluorescence by processes other than photochemistry - the converting of excess excitation into heat.

Nonreducing end : The terminal glucose moiety of the starch molecule whose C-1 is â- linked to the C-4 of the preceding moiety.

Nonreducing sugar : A sugar in which the aldehyde or ketone group is reduced to an alcohol or combined with a similar group on another sugar, e.g., sucrose less reactive than a reducing sugar.

Normal seedling :

- Seedlings that show the capacity for continued growth and development under favourable conditions and possess all essential structures indicating the ability for normal growth and development.
- The emergence and development of essential structures from seed embryo indicating the ability to development into a normal plant under favourable conditions.

Normal solution : Number of gram equivalent weight of a substance in one litre of solution.

Normalized difference vegetation index (NDVI) : Greenness index used to estimate above-ground net primary production from satellites, based on reflectance in the visible and near infrared.

Northern- blot analysis : A method for detecting and quantifying specific RNAs (after electrophoresis and transfer of the RNA onto nitrocellulose paper or a nylon membrane filter) by hybridization with complementary strands of RNA or DNA.

Notified laboratory : Seed testing laboratory approved by the central testing laboratory for testing seed samples meant for certification.

Noxious weed seed (primary) : Seeds of noxious weed species not permitted in a certification crop seed.

Noxious weed seed (secondary) : Seeds of noxious weed species permitted to a limited extent in a crop seed.

NPA (1-N-naphthylphthalamic acid) : A noncompetitive inhibitor of auxin anion efflux carriers that blocks the polar transport of auxin.

Nph1 mutant : Arabidopsis mutant (nonphototropic hypocotyl) whose blue light - dependent phototropism of the hypocotyl is blocked. Also defective in a blue light - stimulated phosphorylation. The mutant nph1 is genetically independent of the hy4 (cry1) mutant. Recently renamed phot1, for phototropin1.

Npq1 mutant : An Arabidopsis mutant (nonphotochemical quenching) defective in the enzyme that converts violaxanthin into zeaxanthin. Lacks a specific stomatal response to blue light.

Nucellus :

- The nutritive tissue surrounding the ovule.
- Upper end of embryo sac.

Nuclear division : Mitosis or meiosis. Distinguished from cytokinesis.

Nuclear endosperm : A type of endosperm in which the early development is characterized by rapid cell enlargement accompanied by nuclear division without cell wall formation.

Nuclear envelope : Double membrane which surrounds nucleus.

Nuclear genome : Genes and their intervening sequences.

Nuclear localization signal : A specific amino acid sequence required for a protein to gain entry into the nucleus.

Nuclear magnetic resonance (NMR) spectroscopy :

- Technique used to make a spectrum of molecules with a permanent magnetic moment, due to nuclear spin; the spectra are made in a strong magnetic field that lines up the nuclear spin in all the molecules; it can, for instance, be used to measure the pH in different cellular

- A nondestructive technique that provides structural information by subjecting the sample to a magnetic field which causes the nuclei of various atoms to orient in one of the two energy levels; the energy absorbed by the molecules lead to change in the energy levels of the molecules which is sensed as a change in ratio frequency; this change in the ratio frequency is measured and reported; this instrument can be used to measure water; oils etc.

Nuclear pore complex : Elaborate structure composed of more than hundred different proteins arranged octagonally.

Nuclear pores : Two membranes of the nuclear envelope join at sites.

Nucleoids : Circular chromosomes, similar to those of bacteria but different from linear chromosomes in the nucleus, localized in specific regions of mitochondrial matrix.

Nucleolar organizer : Portions of one or more chromosomes in nucleolus where ribosomal RNA (rRNA) genes are clustered to form a structure.

Nucleolus : Densely granular region in nuclei, site of ribosome synthesis.

Nucleoside diphosphate kinase 2 (NDPK2) : Protein kinase associated with photochrome.

Nucleosome : Segments of linear double helix of DNA coiled twice around a solid cylinder of eight histone protein molecules.

Nucleus : Organelle that contain the genetic information responsible for regulating metabolism, growth and differentiation of the cell.

Nutrient assimilation : Incorporation of mineral nutrients into organic substances such as pigments, enzymes, cofactors, lipids, nucleic acids, amino acids etc.

Nutrient depletion zone : The region surrounding the root surface showing diminished nutrient concentrations due to slow diffusion.

Nutrient productivity : Rate of plant biomass increment per unit nutrient in the plant.

Nutrient resorption : Withdrawal of nutrients from a plant part during senescence before shedding.

Nutrient solution : A solution containing only inorganic salts that supports the growth of plants in sunlight without soil or organic matter.

Nutrient-use efficiency : Growth per unit of absorbed plant nutrient which equals nutrient productivity times mean residence time of the nutrient.

Nyctinasty : Sleep movements of leaves. Leaves extend horizontally to face the light during the day and fold together vertically at night.

❑❑❑

O

Oat type : Equal distribution on both surfaces. Example - Wheat, Maize, Barley etc.

Objectionable weed seed : Weed seed this is (a) difficult to separate from the crop seed; (b) presence of which can cause genetic contamination in the subsequent stage; (c) parental and difficult to eradicate when established. Such seeds are limited by certification standards.

Obligate anaerobes : Do not require O_2 for respiration. Examples - denitrifying and lactic acid bacteria.

Obligate or qualitative response : An absolute requirement for the proper environmental cues in order to flower exhibited by some plants.

Obligate photoperiodic plants : Species remain permanently vegetative if kept under unfavourable day length. Example - Short day - Xanthium and Tobacco long day - Hyoscyamus niger.

ODV test : Other Distinguishable variety test; A test made on the submitted sample to determine the extent of admixture by other distinguishable variety seeds based on distinct morphological characters of seeds. **e.g.** Mixture of rad rice in white rice.

OECD : Acronym for Organisation for Economic Co-operation and Development. An organization of European countries (presently EEU) that has developed the international seed certification scheme.

Off types : Plants differing in their characteristics from those of seed crop phenotypically or genotypically.

Official seed sample : Seed samples drawn and sent to the seed testing laboratory by seed inspector.

Oils : Fats which are liquid in ordinary temperatures.

Oleosins : Proteins present in half unit membrane help other proteins bind to the organelle surface.

Oleosomes, Lipid bodies or spherosomes :

- Oil containing microbodies.
- Organelles bounded by an unusual single layer phospholipid membrane that store triacylglycerols.

Oligogalacturonans :

- Pectin fragments obtained from degradation of pectin as a result of attack by fungus phytophthora which secretes endo-polygalacturonase.
- Pectin fragments (10 to 13 residues) resulting from plant cell wall degradation that elicit multiple defense responses. May also function during the normal control of cell growth and differentiation.

Oligomycin : Specific inhibitor of the F_0F_1 ATP synthase.

Oligosaccharides : (Greek oligos, "few") Consists of short chains of monosaccharide units joined together by covalent bonds. ex-sucrose or cane sugar.

Oligosaccharins :

- Fragments resulting from plant cell wall degradation that affect plant development or defenses.

- Production of biologically active fragments 10 to 15 residues long resulting from degradation of cell wall that may be involved in natural developmental responses and in defense responses.

Open storage :

- Storage of seeds under ambient conditions of temperature and relative humidity.
- Storage with free access to normal atmospheric conditions.

Operculum : A kind of epistase where cells of integument form a tightly compact plug at the micropyle leading to reduced imbibition.

Opines : Nitrogen-containing compounds found only in crown gall tumours. Genes required for their synthesis found only in Ti plasmid T-DNA of Agrobacterium tumefaciens, which can utilize the opines, not the plant. That bacterium can utilize the opine as a nitrogen source, but cells of higher plants cannot.

Opportunity costs : Diminished growth resulting from diversion of resources from alternative functions that might have yielded greater growth.

Optimum LAI : LAI values beyond which photosynthetic production declines.

Or Nobel (1999) : Sums of all the forces that may act on a molecule to drive net transport.

– or roots, depends on metabolic processes in the guard cells.

Orange seedlot certificate : An international seed analysis certificate provided by accredited seed testing lab for seeds produced in its country.

Ordinary companion cell : A type of companion cell with relatively few plasmodesmata connecting it to any of the surrounding cells other than its associated sieve element.

Ordinary storage : Storage as usually practiced with no attempt to control either temperature or relative humidity of the storage area.

Organ culture - Culture of isolated plant organs e.g. meristem, shoot tip, root culture etc.

Organ identity genes : Genes that directly control floral development. Genes for transcription factors that likely control the expression of other genes whose products are involved in the formation and/or function of floral organs.

Organic fertilizers :

- Contain nutrient elements derived from natural sources without any synthetic additions.
- Originates from the residues of plant or animal life or from natural rock deposits.

Organic nitrogen : Amino acids, amides, urea which have been directly absorbed by some plants,

Organogenesis : A process of differentiation in which plant organs viz. root, shoot, bud, flowers, stem are formed denovo or from pre-existing structures.

orp mutant : In maize, a mutation that inactivates the enzyme for the final step in tryptophan biosynthesis. Prevents conversion of tryptophan to IAA. Named for orange pericarp mutant of maize.

Orthodox seeds : Refers to seeds that can be dried to safe moisure content and stored for reasonable time without loss of viability. **Syn.** Poikilohydrous seeds.

Orthostichy : A set of leaves inserted in the stem directly above or below one another.

Oscillator : An internal pacemaker which governs endogenous nature of circadian rhythms.

Osmolality :

- Moles of total dissolved solutes per litre of water.
- A unit of concentration expressed as moles of total dissolved solutes per liter of water [mal L^{-1}].

Osmophilic : Preferring to grow with very little water.

Osmopriming : An invigouration treatment with controlled imbibition of seed to a level that permits pregerminative metabolism to proceed, but prevents actual emergence of radicle. Syn. Osmotic priming.

Osmoregulation :

- Adjustment of the concentration of osmotic solutes in plant cells in response to changes in soil water potential.
- Control of the osmotic potential of a cell or organism.

Osmosensor : System involved in sensing a change in the concentration of solutes in cells; osmosensors were first extensively studied in yeasts and subsequently also identified in plants.

Osmosis :

- Diffusion of solvent from higher concentration to lower concentration through semipermeable membrane.
- The movement of water across a selectively permeable membrane towards the region of more negative water potential, $Ø_w$ (lower concentration of water).

Osmotic adjustment :

- Net increase in the solute content per cell that is independent of the volume changes that result from loss of water.
- The ability of the cell to accumulate compatible solutes and lower water potential during periods of osmotic stress.

Osmotic potential :

- Measure of decrease in free energy due to solute solvent interaction represented with (-) sign.
- Component of the water potential that is due to the presence of osmotic solutes; its magnitude depends on solute concentration.

Osmotic pressure (ð) :

- The negative of the osmotic potential, ($Ø_s$) Because $Ø_s$ has negative values, ð has positive values.
- Pressure exerted by solutes.

Osmotic priming : Slow hardening of seeds using osmotic chemicals like polyethylene glycol, Mannitol etc.

Osmotic solutes : The major osmotically active solutes in guard cells. Include potassium, chloride, malate, and sucrose.

Osmoticant : A chemical that can create an osmotic effect. E.g. Carbowax, PEG (Poly Ethylene Glycol).

Osmoticum : Substances used to create water or osmotic stress artificially through solution. Examples - Polyethylene glycol 6000, 8000, mannitol etc.

Outer membrane : The outer membrane of the mitochondrion or chloroplast.

Outer space : Part of plant cell or tissues which allows free diffusion to take place.

Ovary : Female sex organ.

Over-expression : The engineering of a specific gene to cause its greater activity than in the nonengineered form.

Ovule :

- In seed plants, the structure that contains the embryo sac and develops into the seed after fertilization of the egg contained within it.
- Port of ovary which forms seeds.
- The structure embedded in the ovary containing ovules, the egg cells surrounded by nucellus and enclosed by integuments the ovule develops into seed after fertilization.

Ovule abortion : The death of ovule in an ovary.

Ovuliferous : Bearing ovules, a term applied to ovule bearing scales in pine cones.

OxIAA (oxindole-3-acetic acid) : End product of oxidative degradation of IAA.

Oxidation :

- Release of electrons or hydrogen from a compound or addition of O_2.
- Act or process of being oxidized.
- The chemical reaction whereby electrons or hydrogen atoms are removed from a substance.

Oxidation number : Number used to keep track of the redistribution of electrons during chemical reactions. The sum of the oxidation numbers of all atoms in a neutral compound is zero. A useful method to balance redox equations.

Oxidative pentose phosphate pathway : Metabolic pathway that oxidizes glucose and generates NADPH for biosynthesis.

Oxidative phosphorylation :

- Redox reactions of the electron transport chain and synthesis of ATP.
- Formation of ATP (from ADP and Pi) coupled to a respiratory electron-transport chain in mitochondria and driven by a proton-motive force.
- Transfer of electrons to oxygen in the mitochondrial electron transport chain that is coupled to ATP synthesis from ADP and phosphate by the ATP synthase.

Oxigenic organism : Photosynthetic organisms that produce molecular oxygen as an end product of photosynthesis.

Oxygenases : Group of enzymes that add oxygen from O_2 directly to organic compounds.

Oxygenation :

- The oxygenase-catalyzed reaction in which the carbon of an organic molecule binds the oxygen atoms of the O_2 molecule.
- The binding of O_2 to a substrate, without changing the redox state of O (e.g., ribulose-1,5-bisphosphate by Rubisco); it also refers to the addition of O_2 to a medium (e.g.,water).

❑❑❑

P

P protein : A set of proteins abundant in the sieve elements of most angiosperms but absent from gymnosperms. Occurs in tubular, fibrillar, granular, and crystalline forms depending on the species and maturity of the cell. Seals off damaged sieve elements by plugging up the sieve plate pores. May play other roles, such as defense against phloem-feeding insects. Formerly called "slime."

P680 : The chlorophyll of the photosystem II reaction center that absorbs maximally at 680 nm in its neutral state. The P stands for pigment.

P700 : The chlorophyll of the photosystem I reaction center that absorbs maximally at 700 nm in its neutral state. The P stands for pigment.

P870 : The reaction center bacteriochlorophyll from purple photosynthetic bacteria that absorbs maximally at 870 nm in its neutral state. The P stands for pigment.

Paclobutrazol (aka : Bonzi) : A commercial inhibitor of gibberellin biosynthesis. Paclobutrazol inhibits P450 mono-oxygenases, blocking the synthesis GA_{12}- aldehyde on the endoplasmic reticulum.

Palea :

- One of the two glumes that occurs in a spikelet (rice). It is smaller than the other glume called lemma.
- Upper bract that with the lemma encloses the flower in grasses.

Paleonophily : Pollination by months. E.g. Tobacco.

Palisade cells :

- Mesophyll cells which are compact mostly found beneath upper epidermis.
- Below the leaf upper epidermis, the top one to three layers of pillar-shaped photosynthetic cells.

Palisade mesophyll : Transversally oriented elongated photosynthetic cells at the adaxial side of a dorsiventral leaf.

Panicle : A branching raceme with each branch bearing flowers in racemose sequence.

Panicle exsertion : Complete emergence of panicle from the boot leaf with visible peduncle.

Paper towel : Germination paper of prescribed quality in roll towel or between paper methods of germination testing.

Paper-Aluminum foil-polythene laminated pouches (P-AF-P) : A moisture vapour proof container made by laminating three different materials for packing seeds for long term storage.

Paper-piercing test : A vigour test to measure the ability of seed to emerge under unfavourable field conditions; conducted by placing the seeds between a layer of sand and filter paper of specified strength followed by a layer of send. Seeds that germinate through this layer are termed vigourous.

Paraheliotropism :

- Leaf movement that positions the leaf more or less parallel to the incident radiation throughout the day.
- Movement of leaves away from incident sunlight.

Paraquat : An herbicide that blocks photosynthetic electron flow by accepting electrons from photosystem.

Parenchyma : Metabolically active plant tissue consisting of thinwalled cells, with air filled spaces at the cell comers.

Parent material : Rock and other substrates that generate soils through weathering.

Parietal cell : Cell that forms along with megaspore mother cell by the division of archesporial cell and normally disintegrates later.

Parthenocarpy : The production of fruits without fertilization, consequently fruits lacking mature functioning seeds occurs naturally in banana and pineapple.

Partial dehydration technique : A method of storing recalcitrant seeds; seeds are dried to critical moisture limit, treated with fungicide and stored in perforated polythene bag. **e.g.** Rubber seeds are stored at 20% moisture and 20°C.

Particle rosettes (terminal complexes) : Large, ordered protein complexexs that are embedded in the plasma membrane and contain cellulose synthase.

Partitioning :

- Differential distribution of photosynthates within the plant.
- The differential distribution of photosynthate to multiple sinks within the plant.

Parts per million (PPM) : Ml or mg of solute dissolved in solvent to make1 litre solution.

Pascal (Pa) :

- Force of one neuton per square meter.
- SI unit of pressure, 1 Pa = 1 kg m^{-1} S^{-2} or 1 Pa = 1 J^{-3}.

Passage cells : Cells of endodermis opposite of xylem without thickenings.

Passive absorption : The spontaneous downhill movement of molecules or ions without involvement of metabolic energy.

Passive transport : Diffusion across a membrane. The spontaneous movement of a solute across a membrane in the direction of a gradient of (electro) chemical potential (from higher to lower potential). Downhill transport.

Pasteur effect : Inhibition of respiration by lack of O_2 and continuation of glycolysis.

Patch-clamp : Electrophysiological method used to study ion pumps and single ion channels.

Pathogenesis-related proteins (PR proteins) :

- Group of proteins produced by pathogen attack; includes hydrolytic enzymes that attack the cell wall of the pathogen, particularly fungi, e.g., glucanases, chitinases.
- Proteins thought to protect plants against pathogens.

Pathogens : Microorganisms capable of infecting other organisms and causing disease.

P-ATPase : An ion pump in which the enzyme protein is phosphorylated by ATP during the catalytic cycle, e.g., the plasma membrane H^+-ATPase.

Pattern realization : Time under which light dependent processes occurs.

Pattern specification : Period under which cells and tissues develop and become competent to react to light (Mohr, 1983).

PCMBS (p-chloromercuribenzene-sulfonic acid) : A reagent that inhibits the transport of sucrose across plasma membranes but does not permeate the cell membrane.

Pcripheral zone : A doughnut shaped region surrounding the central zone in shoot apical meristems consisting of small, actively dividing cells with inconspicuous vacuoles. Leaf primordia are formed in the peripheral zone.

Peak value : Refers to the maximum mean daily germination (cumulative percentage of full seed germination divided by number of days elapsed since sowing date) reached at any time during the test period.

Pectin :

- Cell-wall polymer rich in galacturonic acid.
- Long chain polymers of gatactouronic acids found between two cells of seeds. Mainly composed of pectic acid and propectins and their Ca x Mg salts.

- A heterogeneous group of complex cell wall polysaccharides that form a gel in which the cellulose-hemicellulose network is embedded. Typically contain acidic sugars such as galacturonic acid, and neutral sugars such as rhamnose galactose, and arabinose. Often include calcium as a structural component, allowing extractions from the wall with chelators or dilute acids.

- A hydrated gel phase in which the cellulose - hemicellulose network is embedded.

Pedicel : Stalk of a single flower.

Peduncle : Stalk of an inflorescence.

Pendulous : Hanging down.

Pentosans : Polymers of 5-C sugar molecules found kin seed coats. Example - Mannars.

Pentose phosphate pathway (hexose monophosphate shunt) :

- A cytosolic pathway that oxiddizes glucose and produces NADPH and a number of sugar phosphates.

- Bypass of glycolytic pathway also called hexose mono phosphate. shunt. Pentose sugars are the major intermediate products with production of 36 ATPS.

Percent germination :

- An estimation of planting value of seed lots measured in terms of normal seedlings per 100 seeds produced in the standard germination test.

- Percentage of seeds that produces normal seedlings when tested in a laboratory according to established procedures.

Percent solution : Grams of solutes dissolved in a solvent to make final volume 100 ml which may be V/W, W/W, W/V. Example 5 g of any substance dissolved in 95 ml of solvent will give 5% solution.

Percussion : A method of softening hard seeds by vigourous shaking of seeds in a container. **Syn.** Impaction method.

Perennial : A plant that persists for more than two years with annual flowering behaviour.

Perennial species : Whose individuals typically live more than 2 years; the length of the life cycle can be indeterminate or end after a single seed production event (monocarpic).

Perfect flowers : Flowers with both male (stamens) and female pistils) structures present. **Hermaphroditic flowers.**

Perforation plate : The perforated end walls of vessel elements.

Peribacteroid membrane : Plant-derived membrane that surrounds one or more bacteroids in root nodules.

Pericarp :

- It is the Fruit wall.
- Matured ovulary wall in a seed.
- Seed covering derived from the ovary wall; it may be thin and intimately attached to the seed coat as in a kernel of corn, fleshy as in berries, or hard and dry as in pods and capsules.
- Wall of the ovary comprising of external - epicarp; middle - mesocrap and internal- endcarp; also known as fruit wall.

Periclinal division :

- Cell divisions parallel to the surface of longitudinal axis.
- New cells between daughter cells is in a plane parallel to the closest surface of the plant (parallel to the perimeter). Originated from Greek word Peri (around, surrounding), kline - bed denotes slope or angle of incline.

Pericycle :

- Single layer of parenchymatous cells internal to endodermis.
- Layer of outermost stelar cells, adjacent to the endodermis.

Periderm : Tissue produced by the cork cambium that contributes to the outer bark of stems and roots during secondary growth of woody plants, replacing the epidermis. Also forms over wounds and abscission layers after the shedding of plant parts.

Perinuclear space : The space between the two membranes of the nuclear envelope.

Period : In cyclic (rhythmic) phenomena, the time between comparable points in the repeating cycle, such as peaks of troughs.

Period of a rhythm : Time that elapses between successive peaks or troughs in the cycle.

Peripheral proteins : Protein bound to the membrane surface by covalent bonds, such as ionic bonds or hydrogen bonds and can be seperated from the membrane with high salt solutions or chaotropic agents which break ionic and hydrogen bonds, respectively.

Peripheral zone : A doughtnut shaped region of smaller cells.

Perisperm : Nutritive tissue occurring within seeds derived from nucellus and is diploid e.g. Paper, beet.

Permafrost storage : Storage of germplasm in regions having permanent ice cover or ice caves as in Antarctica. This minimizes the cost on maintaining low temperature and low RH.

Permanent wilting point :

- Soil water potential at which a plant can no longer absorb water from the soil; it is species specific but is generally taken to be -1.5 MPa.
- Water content of the soil (or soil water potential) from which plants cannot regain turgor and therefore remain wilted even if all water loss through transpiration stops.
- Wilting caused by an actual lack of water in soil. A plant will not recover unless water is added to the soil. In most herbaceous plants the leaf water potential is then - 14 to 15 bars.

Peroxidase method : A method of viability/vigour estimation of peroxidase enzyme activity involving the chemical guaiacol, which in the presence of peroxidase enzyme is transformed into blue tetraguaiacoquinone. This chemical is measured calorimetrically.

Peroxidase test : A test for soybean varietal identification. Seeds are soaked in water for 24hours followed by soaking in quaiacol solution (0.05%) for twenty minutes and then placed in 0.1% hydrogen peroxide solution for two minutes; the reddish brown colour obtained is read at 480 nM.

Peroxidation : Formation of peroxides from lipids exposed to O_2 throgh autooxidation or enzyme catalysed oxidation.

Peroxysomes :

- Organelle in which organic substrates are oxidized by O_2. These reactions generate H_2O_2 that is broken down to water by the peroxysomal enzyme catalase.
- Found in eucaryotic organisms in photosynthetic cells involved in removal of hydrogen from organic substrates consuming O_2 in the process forming harmful H_2O_2 which is further broken down in peroxysomes by catalase.
- Microbodies in cells generally associated with chloroplast, contain several oxidative enzymes.

Persistent fruit : A fruit that remains attached to the mother plant after maturity.

Petals : Conspicuous, often brightly coloured flower structures surrounded at their base by sepals.

Petiole : The leaf stalk that joins the leaf blade to the stem.

Pfr : The far-red light-absorbing form of phytochrome converted from Pr by the action of red light. The blue-green coloured pfr is converted back to Pr by far-red light. Pir is the physiologically active form of phytochrome.

pH : Negative of logarithm of hydrogen ion concentration, with rise in pH by one unit hydrogen ion concentration shows ten fold decrease.

Phase : In cyclic (rhythmic) phenomena, any point in the cycle recognizable by its relationship to the rest of the cycle, for example, the maximum and minimum positions.

Phase change :

- The phenomenon in which the fates of the meristematic cells become altered in ways that cause them to produce new types of structures.
- The transition from one phase to another.

Phenol : Compound that contains a hydroxyl group on an aromatic ring.

Phenol test : A varietal identification test used for rice and wheat; soaked seeds are placed on blotter wetted with 1% phenol solution and kept for 3-4hrs. The phenol vapour impart colour on the seed coat due to the formation of melanin pigment formed by polyphenoloxidase enzyme present in the glumes and pericarp. The depth of brown colour is a varietal character and used in classifying the seeds.

Phenol test (modified) : It is similar to phenol test except that seeds are presoaked in copper sulphate or sodium carbonate instead of water.

Phenolic compounds/ Phenolics :

- Compounds that contain a phenol group a hydroxyl functional group on an aromatic ring obtained by shikimic acid pathway or malonoic acid pathway.
- Aromatic hydrocarbons, many of which have antimicrobial and anti-herbivore properties.
- Plant secondary metabolites that contain a hydroxyl group on an aromatic ring, a phenol. Many plant phenolic compcunds function as defenses against insect herbivores and pathogens, UV protectants, pollinator attractants and allelopathic agents.

Phenology :

- The study of flowering and fruiting in plants in relation to seasonal/climate changes.
- Time course of periodic developmental events in an organism that are typically seasonal (e.g., budbreak or flowering).

Phenotype :

- Genotype + Environment +(Genotype x Environment).
- The phenotype characteristics of a plant such as morphological, physiological features governed by the interaction of genotype and environment; such characters are non-transmissible over generations.

Phenotypic plasticity : Range of variation of a trait in a genotype as a result of growth in contrasting environmental conditions.

Phenotypic variability : The total variability which includes genetic and caused by environment observed in a trait.

Phenylalanine ammonia lyase (PAL) :

- Catalyzes the conversion of phenylalanine to form cinnamic acid with loss of an ammonia molecule. An important regulatory step between primary and secondary metabolism.
- Enzyme that catalyzes the fist step in the conversion of the amino acid phenylalanine into phenolics.

Phenylpropanoids : Phenolic dderivauves of chnnamic acid containing a benzene ring and a three carbon side chain.

Pheophytin :

- Colourless chlorophyll a in which central magnesium atom has been replaced by two H atoms.

Phloem :

- Consists of sieve tubes, companion cells and phloem parenchyma.
- Long-distance transport system in plants for mass flow of carbohydrates and other solutes.
- The tissue that transports the products of photosynthesis from mature leaves to areas of growth and storage, including the roots.

Phlem loading :

- The movement of photosynthetic products from the mesophyll chloroplasts to the sieve elements of mature leaves. Includes short-distance transport steps and sieve-element loading.
- Transfer of photosynthates from the mesophyll cells of the leaf to the phloem sieve tube elements.

Phloem unloading :

- Transfer of photosynthates from phloem sieve tube elements to the cells of a sink.
- The movement of photosynthates from the sieve elements to the sink cells that store or metabolize them. Includes sieve-element unloading and short distance transport.

Phosphatase : Enzyme hydrolyzing organic phosphate- containing molecules.

Phosphate : Orthophosphate : inorganic phosphate, HPO_4^{2-},P_{I-}

Phosphate translocator (phostor) : An integral membrane protein that catalyzes the reversible transport of inorganic phosphate in exchange for trioses phosphates (or 3-phosphoglycerate) across the chloroplast envelope. Under illumination, this protein facilitates the increase of trioses phosphates in the cytoplasm while concurrently replenishing inorganic phosphate in the chloroplast for further photosynthetic CO_2 fixation.

3-Phosphoadenosine-5-phospho-sulphate (PAPS) :

- The first stable intermediate in the bacterial and fungal reduction of sulphate (SO_4^{2-}) to Sulphite (SO_3^{2-}) and than sulphide (S^{2-}). Probably formed by reaction of adenosine-5,-phosphosulphate (APS) with ATP. A few plants have more specialized day length requirements or no requirement.
- A crystelline polysaccharide is the major component of cell wall.

Phospholipid :

- Major category of membrane lipids, generally composed of two fatty acids linked through glycerol phosphate to one of a variety of polar groups.
- Integral part of membrane in which they are bound to substances like proteins and carbohydrates.
- Esters of fatty acid and alcohol containing phosphate group.
- Lipids in which two fatty acids are covalently linked to glycerol which itself is covalently linked to a phosphate group.

Phosphorescence : Loss of energy as light during return of electrons of chlorophyll from the higher energy levels to the ground state.

Phosphorylation : Process involving the covalent binding of a phosphate molecule; many enzymes change their catalytic properties when phosphorylated.

Photoassimilates : Product of photosynthesis i.e. carbohydrates.

Photoassimilation : Process of coupling nutrient assimilation to photosynthetic electron transport.

Photoblastic seed : General term for seeds that require light for germination.

Photochemistry : Very rapid chemical reactions in which light energy absorbed by a molecule causes a chemical reaction to occur.

Photochrome B : Phytochrome which mediates response to continuous red or white light.

Photodamage/photodestruction : Damage to/ destruction of components of the photosynthetic apparatus as a result of exposure to high irradiance, frequently in combination with other stress factors; the result is photoinhibition.

Photoinductive cycles : Fixed number of days of requisite light and dark periods.

Photoinhibition :

- Decline in photosynthetic efficiency upon exposure to high irradiance; the decline can be transient (less than 24 hours), which is related to protection of the photosynthetic apparatus, or it can be longer lasting, which implies photodamage.
- Inhibition of germination on exposure to light.
- The inhibition of photosynthesis by excess light.

Photolyase : A blue light-activated enzyme that repairs pyrimidine dimers in DNA that have been damaged by ultraviolet radiation. Contains an FAD and a pterin.

Photolysis of water or Hill reaction : Splitting of water in presence of light to produce H^+ and OH^- ions. H^+ ions are used to reduce CO_2 and OH^- ions recombine to form water, release O_2 and e^-.

$$4H_2 0 = 4\ H^+ + 4\ OH^-, 4\ OH^- = 2\ H_2O + 4e^- + O_2$$

Photometric system for radient energy : Based on sensitivity and brightness response of human eye. Lux meters, Illumination meter or photographic light meters are used for this purpose.

Photomorphogenesis :

- Influence of light on structural development or origin of form.
- Plant development affected by light; generally under control of photoreceptors.
- The influence and specific roles of light on plant development in the seedling light-induced changes in gene expression to support

above-ground growth. In the light rather than below-ground growth in the dark.

Photon :

- Individual particle of light.
- A discrete physical unit of radiant energy.
- Discrete unit of light that describes its particle-like properties (quantum); light also has wavelike properties.

Photon flux density (PFD) : A measure of the level of irradiance in the (near) visible spectral region; it is expressed as photons incident on a (horizontal) plane per unit of time.

Photon Irradiance : Measurement of lightenergy expressed in moles per square meter per second (mol m^{-2} S^{-1}), where 1 mol of light = 6.02 X 10^{23} photons. Avogadro's number.

Photoperiod :

- Length of the daylight period each day.
- Length of the light period on a given day.
- The amount of time per day that a plant is exposed to light or darkness. May control various aspects of sexual or vegetative reproductive development, including flowering and tuberization.

Photoperiodic : Responding to the length of the night.

Photoperiodic induction : The photoperiod regulated processes that occur in leaves resulting in the transmission of a floral stimulus to the shoot apex.

Photoperiodic response : Response by a plant to the duration and order of sequence of light and dark periods. Example - Flowering, internode elongation, seed germination etc.

Photoperiodism :

- A biological response to the length and timing of day and night, making it possible for an event to occur at a particular time of year.
- Response of a plant to the relative length of light and dark periods.
- Response to the timing of light and darkness.

Photophosphorylation :

- The addition of phosphate group to ADP under influence of light energy to form ATP.
- The formation of ATP from ADP and inorganic phosphate (P_i) using light energy stored in the proton gradient across the thylakoid membrane.

Photoprotection : A carotenoid based system for dissipating excess energy absorbed by chlorophyll in order to avoid forming singlet oxygen and damaging pigments. Involves quenching..

Photoreceptor : A protein with chromophore that absorbs light in a specific spectral region; it is typically the start of a signal-transduction pathway leading to photomorphogenetic events.

Photorespiration (C_2 pathway) :

- Increase of respiration rate induced by light under higher light intensities and O_2 with reduced CO_2 concentration, takes place in cooperative way among chloroplast, peroxysomes, and mitochondria. Glycolate being the substrate. Examples – Paddy, Soybean, Wheat etc.
- Production of CO_2 in the metabolic pathway that metabolizes the products of the oxygenation reaction catalyzed by Rubisco.
- Uptake of atmospheric O_2 with a concomitant release of CO_2 by illuminated leaves. Molecular oxygen serves as substrate for rubisco and the formed 2-phosphoglycolate enters the photorespiratory carbon oxidation cycle. The activity of the cycle recovers some of the carbon found in 2-phosphoglycolate, but some is lost to the atmosphere.

Photoreversibility :

- Relating to phytochrome, the interconversion of the Pr and Pfr forms.
- Conversion of Pr. form into Pfr form by absorbing red light and conversion of Pfr to Pr form by absorbing far red light.

$$\text{Pr.} \underset{\text{Far red light}}{\overset{\text{Red light}}{\rightleftarrows}} \text{Pfr.}$$

Photostationary state : Relating to phytochrome under natural light conditions, the equilibrium of 97% Pr and 3% Prf.

Photosynthate : The carbon-containing products of photosynthesis.

Photosynthesis :

- Process in which light energy is used to reduce CO_2 to organic compounds; occurs in chloroplasts in higher plants and algae.
- The conversion of light energy to chemical energy by photosynthetic pigments using water and CO_2 and producing carbohydrates.
- Process by which green plants manufacture complex carbonaceous substances from CO_2 and water in presence of solar energy and chlorophyll . O_2 being the end product.

$$6CO_2 + 12\,H_2O \xrightarrow[\text{Chlorophyll}]{\text{Light}} C_6\,H_{12}\,O_6 + 6H_2O + 6O_2$$

Photosynthetic efficiency : Efficiency of the use of light for photosynthesis (quantum yield); mostly used in conjunction with chlorophyll fluorescence.

Photosynthetic electron transport : Electron flow from light excited chlorophyll and the oxidation of water, through PSII and PS1, to the final electron aceptor $NADP^+$.

Photosynthetic nitrogen-use efficiency : Rate of photosynthesis expressed per unit (organic) nitrogen in the photosynthesizing tissue.

Photosynthetic organs : Organs containing chlorophyll and capable of photosynthesis.

Photosynthetic photon flux density (PPFD) :

- Photosynthetically active radiation expressed on a quantum basis, quanta (mol M^{-2} S^{-1}).
- Refers to the photosynthetically active part of the spectrum.

Photosynthetic quotient : Ratio between CO_2 uptake and O_2 release in photosynthesis.

Photosynthetic units or quantasomes : Smallest group of coordinating pigment molecules necessary to affect photochemical act i.e. absorption and transportation of light quantum to trapping centre where it causes excitation and release of an electron. They contain chlorophylls, carotenoids, quinones, lipids, glycerides and sterols.

Photosynthetic water-use efficiency : Ratio between photosynthetic carbon gain and transpirational water loss.

Photosynthetically active radiation (PAR) or photon flux density :

- Number of photons in the 400-700 nm waveband incident per unit surface area per unit time (mmol.m^{-2}.sec^{-1}) .
- Light of wavelengths between 400 to 700 nm range that corresponds to the wave band absorbed by photosynthetic pigments.
- Radiation used to drive photosynthesis (400–700 nm); the spectral region is similar to that of visible light, but the spectral sensitivity is different from that of the human eye.

Photosystem :

- A functional unit in the chloroplast that harvests light erergy to power electron transfer and to generate a proton motive force used to synthesize ATP.
- Unit comprising pigments and proteins where the excitation energy derived from absorbed photons is transferred to an electron; there are two types of photosystems (I and II) that are embedded in the photosynthetic membrane (thylakoid).

Photosystem I (PSI) : A system of photoreactions that absorbs maximally far-red light (700 nm), oxidizes plastocyanin and reduces ferredoxin.

Photosystem II (PSII) : A system of photoreactions that absorbs maximally far-red light (630 nm), oxidizes water and reduces plastoquinone. Operators very poorly under far-red light.

Phototropins 1 and 2 :

- Protein encoded by phot 1 gene.
- Protein encoded by PHOTi that has been postulated to be a blue-light receptor in phototropism, chloroplast movements and stomatal opening.
- Photoreceptors for blue light signaling pathway.

Phototropism :

- Directional growth towards (or in special circumstances away from) the light.

- Growth of plant organs in response to the directional component of light perceived by the blue-light photoreceptor phototropin.
- The alteration of plant growth patterns in response to the direction of incident radiation.

Phragmoplast :

- A complex assembly of micro tubules, membranes and vesicles that forms during late anaphase or early telophase.
- A complex of microtubules and endoplasmic reticulum.
- An assembly of microtubules, membranes, and vesicles that forms during late anaphase or early telophase and precedes fusion of vesicles to form cell plate.

Phreatophyte : Plant species that accesses deep layers of water.

PHY gene :

- Gene which encodes for phytochrome.
- The genes for the apoproteins of phytochrome : phytochrome gene family members. In Arabidopsis these are PHYA, PHYB, PHYC, PHYD, and PHYE.

Phycobilins : Phycoerythrin (red) and phycocyanin (blue) are available. Involved in transfer of light energy to chlorophyll.

Phyllosphere : Immediate surroundings of a leaf.

Phyllotaxy :

- Arrangement of leaves around stem.
- The arrangement of leaves on the stem.

Phylogenetic constraint : Genetic constitution of a population or taxon that restricts evolutionary change; it can prevent the development of particular traits.

Physical absorption : No involvement of metabolic energy.

Physical maturity : Stage when crop appears dry during visual observations.

Physical scarification : Scarification done using heat (roasting or firing) immersing seeds in boiling water or exposure seeds to microwaves.

Physiological dormancy :

- Dormancy caused by disorganization of functions or of metabolism as brought about by extremes of temperature, oversupply of water, etc.
- Dormancy caused by the presence of physiological inhibiting substances.

Physiological dwarfing : A condition of dwarfing noticed in some coniferous trees that were grown from seeds without stratification. E.g. Embryo extraction technique.

Physiological filter : Physiological limitations due to intolerance of the physical environment, which prevent survivorship of plant species that arrive at a site.

Physiological isolation : Isolation brought about by the altered physiology of the seed crop; this is achieved by planting seed crop at higher altitude, which will result in delayed flowering that will prevent contamination.

Physiological maturity :

- Stage when seeds gain their maximum dry weight.
- A general term used to indicate the full potential of the seed when the seed has highest dry matter accumulation of all biochemical components accompanied by high germination and vigour.

Physiologically dry : No absorption of water by roots despite of sufficient moisture in soil. Examples cold and saline soils or salt marshes.

Phytase :

- An enzyme that catalyes the breakdown of phytin in the seed coat.
- Calcium salt of myo-inositol hexakisphosphate; organic P-storage compound in seeds and endodermis of some plant species and major fraction of organic P in soils.

Phytin : Primary source of phosphorus containing organic salt complexes of Ca, Mg, Mn & K.

Phytoalexins :

- Plant defense compound against microorganism, whose synthesis is triggered by components of microbial origin.
- Chemically diverse group of secondary metabolites with strong antimicrobial activity that accumulate around the site of infection. Examples - isoflavonoides in legumes sesquiterpenes in tomato, potato and bobacco.

Phytoanticipin : Constitutively produced plant defense compound against microorganism.

Phytochelatin : Sulfur-rich peptide which binds (heavy) metals.

Phytochrome :

- A light sensitive, photoreversible blue pigment present in the embryo, responsible for light sensitiveness resulting in dormancy / germination.
- A plant growth regulating photoreceptor protein that absorbs primarily red light and far-red light, but also absorbs blue light. The holoprotein tht contains the chromophore phytochromobilin,
- A protein pigment that absorbs red and far red light most strongly, but that also absorbs blue light.
- Photoreceptor absorbing red or far-red radiation (depending on its configuration); this pigment is involved in the perception of the presence of light, light quality, and day length.
- Photoreceptor pigment controls growth and development in plant occurs as Pr (red) with the light absorption in red region of spectrum at 660 nm and Pfr. (far red) in far red region at 730 nm, respectively.

Phytochrome A : Phytochrome required for the response to continuous far-red light.

Phytochrome - red (Pr) :

- A near stable phytochrome pigment responsive to red wave length (660nM) and presence of this pigment induces dormancy.
- Red absorbing form of photochrome exists in dark grown or etiolated plants.

- Red light-absorbing phytochrome form. This is the form in which phytochrome is assembled. The blue-coloured Pr is converted by red light to the far-red light-absorbing form, Pfr.

Phytochrome controlled seeds : Seeds which specifically require red light to germinate.

Phytochrome kinase substrate 1 (PKS1) : Cytosolic protein.

Phytochrome-far red : A highly unstable phytochrome pigment that initiates germination processes in light sensitive seeds that converts to phytochrome -red form exposure to far red light (720 nM).

Phytochromobilin :

- Chromatophore of phytochrme a linear tetrapyrrole, found in higher plants.
- The linear tetrapyrrole chromophore of phytochrome.

Phytoecdysones : Group of plant steroids that are toxic to insects because of chemical similarity to the insect molting hormone.

Phytoferritin : An iron-protein complex which stores surplus iron in plant cells.

Phytohormone : Plant compound produced in one part of the plant and having its effect in another part at minute concentrations (nanomolar and picomolar range).

Phytol : The long hydrocarbon chain found on the chlorophyll molecule that anchors it to the thylakoid membrane.

Phytomere : A developmental unit consisting of one or more leaves, the mode to which the leaves are attached, the internode below the node, and one or more axillary buds.

Phytomining : Extracting naturally occurring metals from soils, by utilizing the uptake capacity of plants that accumulate these metals.

Phyto-onco-genes : The ipt gene and the two auxin biosynthetic genes which have the ability to induce tumors in plants.

Phytoremediation : Use of green plants to remove, contain, or render harmless environmental contaminants.

Phytosiderophores :

- Iron-chelating organic molecule in grasses.
- A special class of iron chelators produced by grasses and made of amino acids that are not found in proteins.

Piercing technique : A method used for preparing grass seeds whose glume prevents the entry of tetrazolium solution; a sharp needle is used to pierce the glume before the seed is placed for staining.

Pigment systems : Group of pigments efficient for shorter wavelengths i.e. 680 nm (PS II) and longer wave lengths i.e. upto 700 nm (PSI), respectively.

PIN 1 : Membrane protein that forms a part of the complex that transports auxin out of the basal ends of conducting cells during polar auxin transport.

pin1-1 mutant : An Arabidopsis mutant with flowers similar to those of plants treated with auxin transport inhibitors. The inflorescence is impaired in polar auxin transport. Useful in investigating the role of polar auxin transport in floral development.

Pioneer : Species that is a major component of a vegetation at early stages of succession; used in contrast to climax species.

Pit :

- Microscopic regions where the secondary wall of tracheids is absent and the primary wall is thin and porous, facilitating sap movement between one tracheid and the adjacent one.
- Narrow channel through the thick secondary walls of vessel elements in xylem.

Pit membrane :

- Relatively thin structure in each pit which is formed from the primary cell wall and consists of a dense network of hydrophilic cellulose polymers.
- The porous layer between pit pairs, consisting of two thinned primary walls and a middle lamella.

Pit pairs : The adjacent pits of adjoined tracheid cells. A low-resistance path for water movement between tracheids.

Placenta : Any sporangia-bearing surface; in seed plants that part of the carpel bearing ovules.

Planstocyanins : Copper containing proteins bound loosely to inside of thylakoid membrane acts as electron carrier.

Plant cell wall : Thick, cellulosic, permeable, covering surrounding all the plant cells. Cell wall is a non living structure formed of living protoplast.

Plant culture : Culture of seedlings or larger plants.

Plant hormones : Substances that influence plant growth and development at low concentrations. Major classes are abscisic acid, auxin, brassinosteroid , cytokinin. Ethylene, and gibberellin.

Plant molecular vehicle : Recombinant nucleic acid molecule consists of a vehicle part acts as a vector for introducing foreign genetic material in to plant cell and passenger component i.e. genetic material.

Plant tissue culture :

- Cultivation in vitro on nutrient media of plants, seeds, plant parts (tissues, organs, embryos, single cell, protoplast etc.) under aspectic conditions.
- Technique of vegetative propagation that uses single plant cells, undifferentiated group of cells (callus) or organized cell masses (tissue).

Plasma membrane H^+- ATP ase : A P-ATP ase that transports H^+ across the plasma membrane.

Plasma membrane or Plasmalemma : Membrane that serves to separate the cytoplasm from external environment and allows the cell to take up and retain certain substances while excluding others.

Plasmalemma central control model (PCC model) : Proposed mechanism for gravitropism in which calcium channels in the plasma membrane open in response to gravity-induced changes in the distribution of forces exerted by the protoplast, the cytoskeleton, or the cell wall. Contrast with starch statolith hypothesis.

Plasmalemma plasma membrane : External membrane surrounding the cytoplasm.

Plasmid : Circular pieces of extra-chromosomal bacterial DNA that are not essential for the life of the bacterium. Plasmids frequently contain genes that enhance the ability of the bacterium to survive in special environments.

Plasmodesmata (Singular - Plasmodesma) :

- Tubular extension of the plasma membrane, 40 to 50 nm in diameter, that transverse the cell wall and connect the cytoplasms of adjacent cells.
- Microscopic membrane lined channel connecting adjacent cells through the cell wall and filled with cytoplasm and a central rod derived from the ER called the desmotubule. Allows the movement of molecules from cell to cell through the symplast. The pore size can apparently be regulated by globular proteins lining the channel inner surface and the desmotubule to allow particles as large as viruses to pass through.
- Minute membrane-lined channels that traverse the plant cell wall to provide a cytoplasmic pathway for transport of substances between adjacent cells.

Plasmolysis :

- Separation of the cytoplasm from the cell wall due to water loss; only happens in water, not in air.
- Shrinkage of plant cell or tissue in response to outer hypertonic solution.
- Shrinking of the protoplasm of a cell placed in an hypertonic (low osmotic potential) solution, away from the cell wall, due to loss of water.

Plastic biological strain : The organisms function do not return to normal.

Plasticity : The ability of an organism to adjust depending on the external environment.

Plastids :

- Group of double membrane enclosed organelles.

- Refers to cell organelles generally contain pigments viz. chloroplasts, chromoplasts and leucoplasts.

Plastocyanin :

- Small, water soluble copper containing protein that transfers electrons between the cytochrome b_6f complex and P 700.
- A small (10.5 kDa), water- soluble, copper-containing protein that transfers electrons between the cytochrome b_6 f complex and P700. this protein is found in the luminal space.

Plastohydroquinone (PQH_2) :

- The fully reduced form of plastoquinone.
- A small, nonpolar molecule that diffuses readily in the nonpolar core of the thylakoid membrane and is capable of undergoing reduction to plastohydro-quinone. A mobile electron carrier connecting PSII and PSI. Chemically and functionally, similar to ubiquinone in the mitochondrial electron transport chain.
- Quinone compounds act as electron carriers.

Pleated paper method (PP) : A method of germination where seeds are placed in folds (pleats) of a germination paper and kept in a tray; suitable for small seeds.

Pleiotropic : Referring to a gene having more than one (perhaps many) phenotypic effects.

Plenum : An air chamber found in drying unit through which air is sucked into the drier; the velocity of the air is adjusted as per need.

Plumule :

- Major young bud of, the embryo within a seed or seedling from which the aerial part of the plant will develop.
- Part of embryo which forms shoot.

Pneumatophore :

- Specialized portion of the root that emerges from water-logged soils, believed to be used for gas exchange.
- Plant structures that protrude out of the water and provide a gaseous pathway for oxygen diffusion to the roots growing in water or in water-saturated soils.

Poikilohydric : Plants or plant parts (seeds, pollen) that can dry out without losing their capacity to function upon rehydration.

Polar nature of water : Side where hydrogen atoms are attached is electropositive (+) other side where oxygen atom is attached is electronegative (-). Net charge for molecule being neutral.

Pollen : Small structures (microspores) produced by anthers of seed plants contain haploid male nuclei that will fertilize egg in ovule.

Pollination : Transfer of pollen from anther to stigma.

Poly Acrylamide Gel Electrophoresis (PAGE) :

- A technique for varietal identification where electrophoresis is conducted using polyacrylamide gel as the base.
- Standard method for seperation identification and purification of small DNA fragments.

Polycotyl : Seeds with more than two cotyledons. **e.g.** Pinus sp. **Syn.** Polycotyledon, pluricotyledony.

Polycotyledon : Having many cotyledons, **e.g.** Pine seed embryo.

Polymerase chain reaction (PCR) : Technique generates microgram (μg) quantities of DNA copies (up to billion copies) of the desired DNA or (RNA) segment present even as a single copy in the initial preparation in a matter of few hours.

Polymer-trapping model : A model that explains the specific accumulation of sugars in the sieve elements of symplastically loading species.

Polyploid : An individual that carries two or more complete sets of homologous (pairs) chromosome.

Polysaccharide :

- A polymer made of many sugar residues.
- Polysaccharides consist of long chains having 100 or 1000 of monosaccharides units. ex - cellulose.

Polyterpenes ($[C_5]_n$) : High-molecular-weight terpenes containing 1500 to 15,000 isoprene units, e.g., rubber.

Polyterpenoids : Terpenes having more than fifty carbons.

Positional information : The concept that cell position, cell relationships, and associations rather than cell lineage is the important determinant of cell differentiation and tissue formation.

Positive photoblastic seeds : Seeds which require exposure to light for germination . Example - Lettuce, Spruce

Post-illumination CO_2 fixation : CO_2 fixation that occurs briefly after a light pulse.

Potamogeton type : Stomata almost absent if present remain functionless. Example - Submerged plant.

Potassium Dichromate test : A varietal identification test for peas using potassium dichromate solution.

Potassium hydroxide text : A varietal identification test for red rice by soaking seeds in 2% KOH solution. The solution turns red in case of red rice and golden yellow for white rice

Potassium nitrate : A chemical used to break dormancy in seeds.

Potato type stomata : More on under surface than upper surface. Examples - Cabbage, Bean, Pea and Tomato etc.

Poteoglycans : A hybrid macromolecule consisting of an oligo or polysaccharide joined to a polypeptide, the polysaccharide is the major component.

Potometer : Equipment used to record transpiration rate.

ppb part per billion : 1 nmol mol $^{-1}$; 1 ng g $^{-1}$; nl $^{-1}$ (not an acceptable SI unit).

ppm part per million : 1 mmol mol $^{-1}$; 1 mg g $^{-1}$; ml l $^{-1}$ (not an acceptable SI unit).

P-protein bodies : Discrete spheroidal, spindle shaped, or twisted / coiled structures of P-proteins present in the cytosol of immature sieve tube elements. Generally disperse into tubular or fibrillar forms during cell maturation.

Preconditioning : Conditioning of seeds before preparing them for tetrazolium staining. It involves hydration of seeds by soaking or placing in between moist paper towel to soften seeds and also to initiate enzyme activation.

Precursor : Source.

Pre-drying : Adopted for sample where grinding is compulsory for moisture estimation; the samples have to be predried to desired moisture content before grinding.

Pre-germinated seeds : Sowing of germinated seeds (radicle protruded) for early establishment and uniform stand.

Premilk stage : Stage in the development of a seed that precedes the milk stage.

Preprophase band :

- A circular array of microtubules and microfilaments formed in the cortical cytoplasm just prior to cell division that encircles the nucleus and predicts the plane of cytokinesis following mitosis.
- A ring of microtubules encircling the nucleus during mitosis.

Pre-sowing treatments : Refers to all those seed treatments given prior to sowing such as fungicide seed hardening; seed pelleting; bacterial inoculation; scarification; priming **etc.**

Pressure chamber : Chamber in which a plant or part thereof can be pressurized; it is, among others, a part of the equipment used to determine the water potential in the xylem of plant stems.

Pressure potential (Ψ_p) :

- The hydrostatic pressure of a solution in excess of ambient atmospheric pressure.
- Turgor pressure exerted by vacuole in a system, represented with (+) sign..
- Pressure component of the water potential; it is positive in nonplasmolyzed living plant cells (turgor) and negative in the xylem of transpiring plants (suction).

Pressure probe : Microcapillary that is injected in a living cell to measure cell turgor.

Pressure-flow model : A widely accepted model of phloem translocation in angiosperms. It states that transport in the sieve elements is diriven by a pressure gradient between source and sink. The pressure gradient is osmotically generated and results from the loading at the source and unloading at the sink.

Primary active transport :

- Coupled directly to a source of energy other than ATP hydrolysis, an oxidation reduction reaction or the absorption of light by protein carrier (in halobacteria, bacteriorhodopsin).
- The direct coupling of a metabolic energy source such as ATP hydrolysis, oxidation-reduction reaction, or light absorption to active transport by a carrier protein.

Primary cell walls : Thin layer, less than 1 ìm formed in the initial stage of cell formation.

Primary culture : Culture resulting from cells, tissues or organs taken from an organism.

Primary dormancy :

- Seed of many crops fail to germinate immediately after harvesting . Examples - Some varieties of paddy.
- Seeds that are released from the plant in a dormant state. Examples - some varieties of paddy.

Primary elements : Elements required in very huge quantities which includes elements obtained from air CHO and from the soil NPK.

Primary endosperm : The gametophytic tissue found in coniferous seeds that exist before fertilization.

Primary growth : The phase of plant development that gives rise to new organs. Primary growth results from the activity of apical meristems in which cell division is followed by cell enlargement.

Primary inflorescence meristem : Meristem that produces floral meristem on its flanks.

Primary leaves : The first pair of leaves in the seedling other than the cotyledonary leaves.

Primary meristem :

- Meristems that are formed during embryogenesis and are found at the tip of the root and the shoot.
- Root and shoot apical meristems formed during embryo development and seed development.

Primary phloem : Phloem derived from the procambium during growth and development of a vascular plant.

Primary plasmodesmata : Formed during cytokinesis when golgi derived vesicles containing cell wall precursors fuse to form cell plate with numerous pores.

Primary productivity : Rate of plant production over a certain period of time. Also measured in terms of increase in dry weight, organic matter, CO_2 assimilation and solar energy fixed.

Primary response genes ("early genes") : Genes whose expression is necessary for plant morphogenesis and that are expressed rapidly following exposure to a light signal. Often regulated by phytochrome linked activation of transcription factors. Genes whose expression does not require protein synthesis. Expression is independent of protein synthesis.

Primary root :

- First root of a plant that develops from the radicle.
- Root generated directly by growth of the embryonic root or radicle.
- The main root of a seedling developed from the radicle of the embryo.

Primary sample : Each probe of seed taken from a seed bag with a seed trier of hand sampling.

Primary seed : In sugarcane seed sett production, primary seed is the progeny of foundation seed (FS-I) supervised and approved by competent authority.

Primary walls : The first- formed, unspecialized cell walls with similarities in molecular architecture in diverse types of growing plant cells. About 85% polysaccharide and 10% protein by dry weight.

Primary xylem : Xylem derived from the procambium during growth and development of a vascular plant.

Priming : The process of controlled hydration of seeds using osmotic solutions (PEG, Mannitol) so that the seeds will imbibe weter slowly and initiate germination process upto radicle emergence. Seeds prone to imbibition injury are primed prior to sowing.

Primordia : Localized region of the shoot apical meristem characterized by higher cell division, leading to the formation or growth of cells with identifiable future, such as leaf.

pro mutant : In tomato, an ultra-tall constitutive response mutant resulting from the loss of function of a negative regulator.

Procambium :

- Meristem that gives rise to the primary vascular tissue and vascular cambium.
- Primary meristematic cells that differentiate into xytem, phloem, and cambium.

Proembryo :

- A young embryo in its early stages of development.
- Initial form of embryo consists of row of cells in vertical direction formed from zygote.

Progeny : Succeeding generations.

Programmed cell death (PCD) :

Process whereby individual cells activate an intrinsic senescence program accompanied by a distinct set of morphological and biochemical changes, called apoptosis.

Prohexadione (BX-112) : Inhibitor of GA biosynthesis. Inhibits gibberellin 3β-hydroxylase that converts inactive GA_{20} to growth-active GA_{2}.

Prolamellar bodies : Semi crystalline tubular arrays of membrane.

Proline : An amino acid accumulates in leaf cells during water stress causing stomatal closure.

Promeristem : That part of the shoot or root apical meristem containing the stem cells and their immediate derivatives which have not yet begun to differentiate.

Promoter region : Part of DNA that controls neighbouring genes.

Propagules : Initial product of vegetative propagation.

Prophase spindle : Assembling of microtubules at two foci on opposite sides of nucleus.

Proplastid :

- Type of immature, undeveloped plastid found in meristematic tissue that can convert to various specialized plastid types, such as chloroplasts, amyloplasts, and chromoplasts, during development.
- Meristem cells contain, have few or no internal membranes, no chlorophyll, and an incomplete complement of enzymes for photosynthesis.

Prosthetic group : A metal ion or small carbon compound (other than an amino acid) that is covalently bound to a protein and is essential for its function.

Protease/proteinase : Protein-hydrolyzing enzyme.

Proteasome : Large proteolytic complex that degrades the cyclins in the cytosol.

Protein bodies :

- Specialized protein storing vacuoles abundant in seeds.
- Protein storage organelles enclosed by a single unit membrane; found mainly in seed tissues.

Protein kinases : Enzymes that have the capacity to transfer phosphate groups from ATP to specific amino acid, such as histidine, serine, threonine, or tyrosine, located either within themselves or on other proteins. Play important roles in enzyme regulation, gene expression, and signal transduction.

Protein phosphatases : Enzymes that remove regulatory phosphate groups from proteins. Have important roles in enzyme regulation, gene expression, and signal transduction.

Protein turnover : Breakdown and synthesis of proteins that does not account for a change in protein concentration.

Proteoid root (¼cluster root) cluster root : A shortlived dense package of root hairs that exudes nutrient solubilizing compounds; the name stems from the family of the Proteaceae.

Protocarnivory : Capability of plants to digest arthropods or other organic items that are trapped on sticky surfaces or in "tank" traps and absorb the breakdown products of the trapped material.

Protochlorophyllide : Etioplast contains a pale yellow precursor pigment.

Protochlorophyllide "a" : A pigment that absorbs red and blue light and becomes reduced to chlorophyll a.

Protoderm :

- In the plant embryo, the surface layer one cell thick that covers both halves of the embryo and will generate the epidermis.
- Meristem which gives rise to the epidermis.

Protomeristem : An embryonic structure that will become the root or shoot meristem upon germination.

Proton co-transport : Transport mechanism that allows movement of a compound against the electrochemical gradient for that molecule, using the proton-motive force.

Proton motive force (PMF, ?P) :

- Gradient of electrochemical potential for H^+ across a membrane, expressed in units of electrical potential.
- Driving force across cell membranes due to a membrane potential and/ or proton gradient.

Protonema (plural protonemata) : An algal-like filament of moss cells generated by germination of moss spores. Protoplasm : Classically, the living substance of all cells. The plasma membrane and all the active organelles, substances, and processes required for life and contained within the plasma membrane.

Protoplasm : Essential, complex, living substance of cells on which all the vital functions of nutrition, secretion, growth, and reproduction depend.

Protoplasmic streaming : Flow of the cytoplasm, mediated by the cytoskeleton.

Protoplast :

- The living contents of a cell enclosed by the plasma membrane : the cytosol, organelles, and nucleus. What remains after removal of the cell wall.
- All components of plant cell excluding cell wall.

- Cell membrane with cytoplasm and cell organelles inside; it is isolated after enzymatic removal of the cell wall.

Protoplast culture : Culture of plant protoplast.

Protoplast fusion : Fusion of cells without cell wall.

Provenance :

- Origin, source, place where found.
- Refers to the area or region or origin of the seeds or trees.

Pseudocyclic photophosphorylation : Illuminated chloroplasts produce ATP in absence of CO_2 and $NADP^+$ but in presence of FMN or Vitamin K and O_2.

Pteridine : Nitrogen-containing compound composed of two six-membered rings. Component of riboflavin and parent compound of pterins.

Pterins :

- A carbon compound to which molybdenum is complexed, forming a prosthetic group in the nitrate reductases of higher plants. also a blue light absorbing chromophore in the DNA repair enzyme photolyase. Chemically derived from pteridines.
- Light absorbing, pteridine derivates that often function as pigments in insects, fishes and birds.

Pulvinus (plural pulvini) :

- A turgor-driven organ found at the junction between the blade and the petiole of the leaf that provides a mechanical force for leaf movements.
- "Joint" in a petiole that allows the movement of a leaf, due to transport of ions between cells in the pulvinus, followed by changes in turgor (e.g., in many legumes).

Pumps :

- Membrane proteins that carry out primary active transport across a biological membrane. Most pumps transportions, such as H^+ or Ca^{2+}.
- The membrane protein that carryout primary active tranport.

Pure Live Seed (PLS) : An index of the quality of seed lots got by multiplying purity (%) X Germination (%).

Pure pellets : Categorisation of pelleted seeds during purity analysis as **(a)** entire pellets in which more than half the surface of the seed is covered by pelleting material except when it is quite sure that the seed is not of the species as stated by the sender or **(b)** When the seed is not of the species as stated by the sender or **(c)** When no seed is present.

Purines and pyrimidines : Nitrogenous basis in nucleic acids. Adenine and guanine are purine, whereas, thymine, cytosine and uracil are pyrimidine basis.

Purity analysis : A test to determine the level of contamination of the seed sample.

Purity percent : The ratio of pure seed weight to the working sample weight expressed in percentage; the purity percent is fixed for all crops by the certification agency.

Purity work board : Small wooden boards with smooth white surface and an area of 150-225 sq. inches used for purity analysis.

Purple seed : Soybean infected by seed borne pathogen Cercospora kkuchi, that imparts purple stain.

Pyrethroids : Monoterpene esters with high insect toxicity. Both natural and synthetic forms are popular ingredients in commercial insecticides.

Pyridoxal phosphate (vitamin B6) : A cofactor required by all transamination reactions.

Pyrophosphate : Two phosphate groups linked by a phosphate ester bond, PP_i

❑❑❑

Q

Q 10 (Temperature coefficient) :

- Ratio of rate of biochemical reaction at a given temperature and ten degrees lower.
- Change in rate of a reaction in response to a 10°C change in temperature.

Q cycle : A mechanism of electron and proton flow in which plastohydroquinone is oxidized, with one of the two electrons passed along a linear electron transport chain towards photosynstem I, while the other electron goes through a cyclic process that increases the number of protons pumped across the membrane.

Qualitative defense compound : Highly toxic secondary plant metabolite that protects against attack by herbivores at low concentration.

Qualitative long-day plant : Plant that will not flower unless the length of the night gets below a critical value.

Qualitative short-day plant : Plant that will not flower unless the length of the night gets above a critical value.

Quality seed : Seeds possessing qualities such as genetic purity; physical purity; vigour and seed health.

Quantitative defense compound : Secondary plant metabolite that gives some protection against attack against a broad range of herbivores when present in large amounts.

Quantitative long-day plant : Plant whose flower induction is promoted by exposure to short nights.

Quantitative short-day plant : Plant whose flower induction is promoted by exposure to long nights.

Quantum (plural quanta) :

- A discrete packet of energy contained in a photon.
- Photon containing the energy.

Quantum efficiency : Photosynthetic yield per photon flux of absorbed light.

Quantum flux density : A measure of the level of irradiance; it is expressed as quanta incident on a (horizontal) plane per unit of time.

Quantum requirement : No. of light quanta required to reduce one molecule of CO_2 and release of one molecule of O_2. 8 light quanta are required for this process. Therefore quantum yield is 1/8 or 12% (Emerson and Lewis, 1943).

Quantum yield :

- Moles of CO_2 fixed or O_2 evolved in photosynthesis, or electrons transported in the photosynthetic membrane, per mole of quanta absorbed; in the context of gas exchange often restricted to the linear, light-limited part of the photosynthesis–irradiance curve; when measuring chlorophyll fluorescence, it refers to the full range of photosynthetic irradiance.
- The ratio of the yield of a particular product of a photochemical process to the total rumber of quanta absorbed.

Quarantine : The act of preventing entry by detecting entry of exotic plants pathogens, pests from other countries by detecting and taking adequate control measures to ensure their destruction.

Quenching : The process by which energy stored in light excited chlorophylls is rapidly dissipated mainly by excitation transfer or photochemistry.

Quercetin : An endogenous flavonoid compound that may inhibit and regulate auxin transport by blocking auxin efflux.

Quiescence : A phenomenon where a non dormant seed remains non germinable.

Quiescent center :

- Central region of root meristem where cells divide more slowly than surrounding cells, or do not divides at all.
- Region of root apex with slow cell division rate.

Quinone : A small, nonpolar molecule that diffuses readily in the nonpolar core of the membrane and is capable of undergoing reduction to hydroquinone.

❑❑❑

R

R genes : Resistance genes that function in plant defense against fungi, bacteria, and nematodes in some cases by encoding protein receptors that bind to specific pathogen molecules, elicitors.

Rachis : Main axis of an inflorescence.

Radial axis : The pattern of concentric tissues extending from the outside of a root or stem into its center.

Radial patterning : Arrangement of tissues a radial pattern in stems and roots extending from the outside of a stem or a root into its center.

Radiant energy : The energy of the sun is derived from nuclear reaction takes place at higher temperature and pressure involving the conversion of hydrogen atoms in to helium.

Radicle :

- Part of embryo which forms root.

- Rudimentary root or lower end of the hypocotyl of the embryo; it forms the primary root of the young plant.
- The embryonic root. Usually the first organ to emerge on germination.
- The root initial of the embryo developing into the primary root after emergence through the seed coat during germination.

Radioactive isotopes : Isotopes which emit radiation (Alpha, Beta or Gamma) when disintegrate. Examples - C^{14} which emit B radiation on disintegration.

Raducing sugar. : A sugar with an aldehyde or ketone group available for oxidation, e.g. glucose, fructose.

Rate of germination : A vigour estimate based on rapidity of seed germination calculated as follows.

$$= \frac{\text{No. of Seedlings emerged}}{\text{Day 1}} + \frac{\text{No. of Seedlings emerged}}{\text{Day N}}$$

Ratio of red light (R) to far red light (FR) : Defined as follows :

$$R/FR = \frac{\text{Photon fluence rate in 10 nm band centred on 660 nm}}{\text{Photon fluence rate in 10 nm band centered on 730 nm}}$$

Raw seed : Harvested seeds that are yet to be cleaned and graded. **Syn.** Field run seed.

Ray stem cells : Small cells whose derivatives include the radially oriented files of parenchyma cells known as rays.

rbes gene : Gene required for activation of Ribulose biphosphate carboxylase enzyme.

rcn1 mutant : Arabidopsis (roots curl in NPA) mutant that shows high sensitivity to NPA, inhibition of hypocotyls elongation, and auxin effiux. RCN1 gene is closely related to the regulatory subunit of protein phosphatase 2A, a serine/threonine phosphatase.

Reaction center complex : A group of electron transfer proteins that receive energy from the antenna complex and convert it to chemical energy using oxidation reduction reactions.

Reactive oxygen species : Toxic forms of oxygen, including the superoxide anion (O_2^-), hydrogen peroxide (H_2O_2) and the hydroxyl radical (.OH). Generated by cells in the vicinity of infection and preceding the hypersensitive response.

Real value of seed : The percentage by weight of seed, which will produce seedlings of the variety under certification.

Recalcitrant organic matter : Soil organic matter that takes a long time to be decomposed.

Recalcitrant seeds :

- Refers to the storage behaviour of those seeds that loose viability when dried below a critical moisture level (20 - 50% of wet weight basis) and are thermo sensitive. e.g. Mango, jack etc.
- Seeds that do not tolerate desiccation and are consequently difficult to store for longer periods; they typically germinate shortly after dispersal without first going through a phase of dormancy.

Receptacle : The basal portion of the thalamus often enlarged or elongated along with the floral parts.

Receptor ligand : Affinity of a receptor domain for signaling molecule, often a small protein or peptide.

Receptor : Protein with a high affinity and specificity for a signaling molecule (e.g., a phytohormone), which is the start of a signal-transduction pathway.

Reclamation : Chlorotic leaf tip and failure to set seeds caused by Cu deficiency

Red drop phenomenon : Dropping of quantum yield near far red region of the spectrum begins at wavelengths greater than 680 nm in green plants and 650 in red algae.

Redifferentiation : The ability of component cells of callus to form a whole plant.

Redox potential : Phenomenon of release or acceptance of electrons or ratio between the reduced and oxidised forms of a compound.

Redox reactions : Chemical reactions involving simultaneous oxidation and reduction of molecular species.

Reduction :

- Addition of electrons or hydrogen to a compound.

- The chemical process by which electrons or hydrogen atoms are added to a substance.

Reductive pentose phosphate pathway : Metabolic pathway that utilizes NADPH produced in the light reaction of photosynthesis and produces triose-phosphate.

Reflectance : Fraction of radiation incident on a surface that is reflected (e.g., a leaf, or the Earth surface).

Rejuvenation :

- Reversion to juvenility.
- The reversion of adult shoots to juvenile shoots. May be promoted by hormones, mineral deficiencies, low light, water stress, defoliation, and low temperature.

Relative elemental growth rate : Fractional change in length per unit time.

Relative growth rate (RGR) : Rate of increase in biomass per unit of biomass present. (Blackman, 1919; Fisher 1921).

$$RGR = \frac{Ln\,W_2 - Ln\,W_1}{t_2 - t_1}\ \text{Weight weight}^{-1}\ \text{Time}^{-1}$$

W_1 & W_1 are dry weight of plants at two successive intervals t_1 & t_2, Ln - Natural log.

Relative humidity (RH) :

- Moisture required to saturate atmosphere at a given temperature.
- Ratio of the quantity of water vapour actually in the air to the greatest amount possible at a given temperature.
- Water vapour concentration of air relative to the maximum water vapour concentration at that temperature.
- Ratio of quantity of water vapour that can be held in a given volume of air at particular temperature; expressed in per cent.

Relative water content : Water content of a plant tissue relative to the water content at full hydration.

Reporter gene : A gene whose expression conspicuously reveals the activity of another gene. Gene engineered to share the same promoter as another gene.

Repressors : Proteins that either alone or in concert with other proteins repress expression of a gene.

Reserve formation : Build-up of storage products that result from diversion of plant resources to storage from alternative allocations, such as growth.

Resistance (against stress) : Plant capacity to minimize the impact of stress factors in the environment, either by the presence of tolerance mechanisms or by avoidance of the stress.

Resistant auxotrophic double mutants : Selection of one parent which possess the resistance against the antibiotics etc. from many potential fusion partners of wild type constitution.

Resonance transfer : The nonradiative, molecule to molecule transfer of energy from one excited molecule to another, such as the transfer on energy from an antenna complex to the reaction center.

Resorption : Translocation of nutrients and soluble organic compounds from senescing tissues prior to abscission.

Resource competition : Use of the same pool of growth-limiting resources by two or more plants.

Respiration :

- Oxidation-reduction process occurring in all living cells whereby chemical action occurs, producing compounds and releasing energy that are partly used in various life processes.
- Process of energy release by oxidation of food (substrate) material. Letin word Respirare (breathe).
- The complete oxidation of carbon compounds to CO_2 and H_2O, using oxygen as the final electron acceptor. Energy is released and conserved as ATP.

Respiration and breathing : Respiration is a chemical process occurs in all organisms, whereas, breathing is a physical proces occurs in few organisms.

Respiratory climacteric : Initial higher respiration rate in fruits kept for storage followed by a decline. Fruits are called as climacteric fruits. Examples - Apple, Apricot, Fig, Mango, Muskmelon, Papaya, Peach, Plum, Tomato, Watermelon etc.

Respiratory control ratio : The ratio of oxygen-consumption rates of isolated mitochondria in the presence and absence of ADP. It is used as a measure of the quality of a mitochondrial preparation.

Respiratory inhibitors : Substances which inhibit respiratory pathways. Examples - Cyanide, azide, carbon mono - oxide, sodium fluoride, oxalo acetate etc.

Respiratory quotient (RQ) : Ratio of CO_2 evolved in respiration to O_2 consumed in the process per unit time. It is 1.0 for carbohydrates, less than 1.0 in fats and proteins and more than 1.0 for organic acids.

Respiratory substrate : Energy rich material used for purpose of oxidation.

Respiroscope : Equipment used to determine respiration rate.

Restriction Fragment Length Polymorphism (RFLP) : Variations occurring within a species in the length of DNA fragments generated by specific endonuclease enzyme; used in varietal identification.

Resurrection plant : Plant that withstands complete dehydration and resumes functioning upon rehydration.

Retarded root : A form of abnormal seedling with a short and weak root that is not in balance with other parts.

Rheological properties : Referring to the flow properties of a material.

Rhizobia :

- Symbiosis occurs between members of family, leguminosae and soil bacteria of genera Azorhizobium, Brady rhizobium, Phytorhizobium, Rhizobium and Sinorhizobium collectively called Rhizobia.
- Collective term for bacteria that fix N_2 in symbiosis with legumes or Parasponia of the genera Rhizobium, Bradyrhizobium, Sinorhizobium, Mesorhizobium, and Azorhizobium.
- Collective term for the genera of soil bacteria that form symbiotic (mutualistic) relationships with members of the plant family Leguminosae.

Rhizome : A horizontally developed underground stem that is used for vegetative propagation of crops like ginger.

Rhizosphere :

- Microenvironment surrounding the root.
- The immediate microenvironment surrounding the root.
- Zone of soil influenced by the presence of a root.

Rhizotrons : Boxes 240 X115 X 20 cm generally with one broad side made of a removable transparent plexiglass with packed soils having bulk density of .28 cm 3 used to study root growth.

Rhythms period : The time between comparable points in the repeating cycle. The period is measured as the time between consecutive maxima (peaks) or minima (troughs).

Rhythms phase : Any point in the cycle that is recognisable by its relationship to the rest of the cycle. the most obvious phase point are the peak and trough positions.

Rib meristem zone : Meristematic cells beneath the central zone that give rise to the internal tissues of the stem in shoot apical meristems.

Riboflavin : A vitamin that is part of FAD and FMN.

Riboside :

- In which a ribose sugar attached to the 9 nitrogen of the purine ring.
- In which the ribose sugar moiety contains a phosphate group.

Rieske iron- sulphur protein : A protein in which two iron atoms are bridged by two sulphur atoms, with two histidine and two cysteine ligands.

Riffle type divider : A open end divider, with channels / spaces found at the top. There are minimum 20 channels / spaces and a larger spade like ladle is used to pour seeds. Seeds are divided by channels/spaces and are collected at the base that has two receptacles.

Ring : Porous wood in which xylem vessels produced early in the growing season are longer and wider than those produced in late wood, adding to the distinction of annual growth rings.

Rogue : A plant that deviates from seed crop genetically and phenotypically; also refers to plants with designated diseases.

Rock phosphate : Inorganic phosphate compound with very low solubility.

Roguing : A general term used for the removal of offtypes, rogues, volunteer plants, diseased plants, designated weeds etc. from a seed crop.

Roguing space : The space provided in a mixed planted rice seed crop to walk through during field inspection and while roguing. A roguing space of one foot is provided forever 6′ for foundation crop and 8′ for certified seed crop.

Roll towel : A method of germinating seed in between two paper towels.

Room germinator : A room converted into a germinator provided with stands for placing germination trays; tube lights; humidifier to control the relative humidity and air conditioner to maintain controlled conditions for germination.

Root axis : Principal line along which a root may grow; hypothetical central line of a root.

Root cap : Cells at the root apex that cover and protect the meristematic cells from mechanical injury as the root moves trough the soil. Site for the perception of gravity and signaling for the gravitropic response.

Root cap stem cells : Meristematic cells that give rise to the root cap.

Root cap-epidermal stem cells : Generate the epidermis of the root cap by anticlinal cell divisions and generate the leteral root cap by periclinal divisions followed by anticlinal divisions.

Root density : Total root length per unit soil Volume.

Root hair :

- Fine elongation of unicellular root epidermal cells involved in water and mineral absorption from the medium.
- Microscopic extensions of root epidermal cells that greatly increase the surface area of the root, thus providing greater capacity for absorption of soil ions and, to a lesser extent, soil water.

Root hair zone : Part of root covered with root hairs associated with higher absorption rate.

Root hydraulic conductance : Water movement through the root is to consider the root as a single pathway having a single hydraulic conductance.

Root pressure :

- Corticle cells of roots forces the water up to the xylem with certain force.
- A positive hydrostatic pressure in the xylem of roots.
- Positive water potential in the xylem due to ion transport into the xylem of roots and subsequent osmotic uptake of water.

Root shoot ratio : Ratio between root and shoot Biomass.

Root turnover : Replacement of (old) roots by new ones, not accounting for a change in the total amount of roots.

Root-mass ratio (RMR) : Ratio between root biomass and total plant biomass, synonym is root mass fraction (RMF).

Rotenone : Specific inhibitor of complex I.

Roughness : Unevenness of a surface that creates turbulence and enhances convective exchange between the surface and the atmosphere.

R-type channels : A type of gated channels for anions that opens or closes very rapidly in response to a voltage change.

Rubisco activase :

- Protein catalyzing the carbamylation of Rubisco that regulates its activity; chaperone protein protecting the catalytic sites of Rubisco at extreme temperatures and in Darkness.
- An enzyme that facilitates the dissociation of sugar biphosphates-rubisco complexes and in so doing, activities rubisco.

Rubisco (ribulose-1,5-bisphosphate carboxylase/ Oxygenase) :

- Enzyme catalyzing the primary step in the Calvin-cycle, the attachment of CO_2 to the CO_2-acceptor molecule ribulose 1,5-bisphosphate (RuBP); also catalyzes the oxygenation of RuBP.

- The acronym for the chloroplast enzyme ribulose bisphosphate carboxylase / oxygenase. In a carboxylase reaction, rubisco uses atmospheric CO_2 and ribulose-1,5- bisphosphate to form twc molecules of 3-phosphoglycerate. It also functions as an oxygenase that incorporates O_2 to ribulose-1,5 -bisphosphate to yield one molecule of 3-phosphoglycerate and another of 2-2-phosphoglycolate. The competition between CO_2 and O_2 for ribulose-1,5-bisphosphate limits net CO_2 fixation.

RUB_S : A family of small, ubiquitin-related proteins. Proteins linked to RUB are usually activated rather than degraded.

Ruderal species : Species that flourish on disturbed sites and complete their life cycle relatively rapidly.

Rudimentary embryo dormancy : A type of dormancy caused by immature embryo at the time of seed shedding. Such types occur in *Ranunculus, Plantago, Fraxinus, Vibumum and Pinuis.*

Ruminate endosperm : The condition of the endosperm that has intruded into the embryo exhibiting irregular structure on the surface and in cross section. e.g. Arecanut.

Run away water : After a heavy rainfall or irrigation some of water drains away along slopes,not available for plant growth.

Runoff : Gravitational water loss from an ecosystem; the difference between precipitation and evapotranspiration (surface and groundwater runoff).

□□□

S

S- gene : A gene conferring the trait of self-incompatibility.

SAG genes : Senescence-associated genes whose expression is induced during senescence.

Salicyl hydroxamic acid : Specific inhibitor of the alternative oxidase.

Salicylic acid : A benzoic acid derivative believed to be an endogenous signal for SAR.

Saline soils : Soils with high salt concentration.

Salinity :

- High concentrations of total salts in soil.,
- Refers to high concentrations of total salts in the soil contrast with sodicity.

Salinization : The accumulation of mineral ions, particularly sodium chloride and sodium sulphate, in soil often due to irrigation.

Salt gland : Group of cells involved in salt excretion.

Salt Stress : The adverse effects of excess minerals on plants.

Salt-tolerant plants : Plants that can survive or even thrive in high-salt soils.

Sampling intensity : The prescribed minimum number of primary samples to be drawn from a seedlot either in containers or in bulk to get the required quantity of composite sample.

Sand benches : A concrete ledge in the germination room filled with sand (upto 5cm depth) used for germination.

Sand media : A medium for germinating seeds containing quartz or sand particles of the size between 0.05 to 0.8 mm in diameter; it must be free of organic matter and sterilized to kill microorganisms.

Sand method : A method of germinating seeds in sand medium by placing the seeds at a depth of 1-2 cm depending on seed size.

Sap : Fluid content of the xylem, sieve elements of the phloem and the cell vacuole.

Saponification : Hydrolysis of fats to yield fatty acids and glycerol.

Saponins :

- A group of triterpenes thought to disrupt fungal membranes by binding to sterols. Mutant oat lines with reduced saponin levels had much less resistance to fungal pathogens than wild type oat (Papadopoulou et al. 1999).
- Secondary plant compound with soaplike Properties.
- Toxic glycosides of steroids and triterpenes with detergent properties. They may interfere with sterol uptake from the digestive system or disrupt cell membranes.

Sapwood : Most recent wood in the xylem of a tree trunk, with open xylem conduits that still function in water transport; it has often a lighter color than the innermost heartwood.

Satellite RNA : Transcribed from DNA led to reduction in disease symptoms.

Saturated fatty acids : Single bond between carbon atoms, most abundant in palm oil. Examples - Palmitic acid ($C_{15}H_{31}COOH$) and steraric acid ($C_{15}H_{35}COOH$).

Saturated solution : Maximum quantity of solute dissolved in a given quantity of solvent under given temperature and pressure.

Saturation : Refers to a condition under which an increase in a stimulus does not elicit a further increase in a response. A maximum state; not capable of further increase, movement, or inclusion.

Saturation deficit : Difference between amount of moisture present in air and moisture required to saturate atmosphere.

SAUR genes : In soybean, a group of primary response genes stimulated by auxin within 2 to 5 minutes of treatment.

SAURs (Small auxin up-regulated RNAs) : Gravitropism which leads to a rapid asymmetry in the accumulation of group of auxin stimulated in RNAs in soybean hypocotyls.

Scarification :

- Any treatment physical or chemical that weaknes the seed coat to overcome dormancy. Method is applied for crops or varieties of hard seed coat Examples - Legumes.
- Breaking, scratching, or softening the seed coat to allow moisture penetration.
- The process of making the hard impermeable seed coat, soft and permeable to water and gases by rubbing; using acid, alkali or hot water.

Scarified seed : Seed that has had its coat scratched to permit ready entry of water and exchange of gases.

Scarifier-huller : A machine that scarifies hard seeds like lucerne; sweet clover by scrubbing action.

Schizocarp : A fruit derived from a simple or compound ovary in which the locule separates at maturity to form simple seeded units.

Schizocotyl : Cotyledons of seedling split longitudinally.

Schizogenous aerenchyma : Gas-transport tissue in plants that is the outcome of highly regulated and species-specific patterns of cell separation and differential cell expansion that creates spaces between cells.

Sclereid : A type of nonelongated sclerenchyma cell commonly found in hard structures such as seed coats.

Sclerenchyma :

- Plant tissue composed of cells, often dead at maturity, with thick, lignified secondary cell walls. It functions in supports of non-growing regions of the plant.
- Tissue that can consist of two types of cells : sclereids and fibers, which both have thick secondary walls and are frequently dead at maturity.

Scleromorph : Containing a relatively large amount of sclerenchyma.

Sclerophyllous leaves : That are scleromorph; they are thick, tough and have a thick cuticle.

Scorable markers : Markers identified in regenerated shoots tested for expression of marker gene.

SCR gene : Arabidopsis SCARE-CROW gene that controls tissue organization and cell differentiation in the embryo, hypocotyls, primary roots, and secondary roots.

scr mutant : Arabidopsis mutant in which hypocotyls and inflorescence are agravitropic and lack both endodermis and starch sheath.

Scutellum :

- Is the single cotyledon in a monocot seed surrounding the embryonic axis and serves as a passage for nutrients from endosperm to embryo.
- Specially differentiated cotyledon in grass seeds; shield shaped organ through which the embryo absorbs food from the endosperm.
- The single cotyledon of the grass embryo, specialized for nutrient absorption from the endosperm.

Sealed storage : Storage in a sealed container; usually refers to hermetic storage.

Second messenger : Intracellular molecule (e.g., cyclic AMP, cyclic GMP, calcium, IP_3 or diacylglycerol) whose production has been elicited by a systemic hormone (the primary messenger) binding to a receptor (often on the plasma membrane). Diffuses intracellularly to the larget enzyme or intracellular receptor to produce and amplify the response.

Secondary active transport :

- Active transport that uses energy stored in the proton motive force or other ion gradient, and operates by symport or antiport.
- Transport involves proton motive force generated by electrogenic H^+ to drive the transport of many other substances against their gradient of electrochemical potentials.

Secondary cell walls : Thicker and stronger and are deposited when most cell enlargement has ended.

Secondary dormancy :

- Seeds that are released from the plant in a nondormant state, but that become dormant if the conditions for germination are unfavourable. Example - Seeds of oat.
- Some seeds which are capable of germinating as soon as harvested lose this capacity if kept in an unfavourable conditions which includes higher temperature, moisture and O_2 content .

Secondary elements : Major elements which are required in less quantities than primary elements. Example - Ca, Mg, S.

Secondary growth : Growth which involves two lateral meristems, vascular cambium gives rise to secondary xylem (wood) and cork cambium which produce secondary phloern.

Secondary inflorescence meristem : Their activity repeat the pattern of development of the primary inflorescence meristem.

Secondary meristems :

- Meristems that are formed after seed germination and include axillary meristems and lateral meristems. Their activity may be suppressed by active primary meristem.
- Vascular cambium and meristematic zones of monocot nodes and grass leaves.

Secondary metabolites (secondary products) :

- Plant compounds that have no direct role in plant growth and development, but function as defenses against herbivores and microbial infection by microbial pathogens, attractants for pollinators and seed-dispersing animals, and as agents of plant-plant competition.
- Products that are large, diverse array of organic compounds that appear to have no direct function in growth and development also known as secondary products or natural products, generally found in only one plant species or related group of species. They protect plants against being eaten by herbivores and against being infected by microbial pathogens. Also serve as attractants for pollinators and seed dispersing animals and as agents of plant - plant competition. Divided into three chemically distinct groups viz. terpenes, phenolics and nitrogen containing compounds.
- Cell constituents which are not essential for survival. Examples - Alkaloids, glycosides (steroids and phenolics), terpenoides and a variety of flavours, perfumes, agrochemicals etc.
- Compounds produced by plants that are not essential for normal growth and development; they are frequently involved in the interaction with a plant's biotic and abiotic environment.

Secondary plasmodesmata : Formed between cells after their cell walls have been deposited. They arise either by evagination of the plasma membrane at the cell surface or by branching from a primary plasmodesmata (Lucas and Wolf, 1993).

Secondary response genes ("late genes") :

- Genes whose expression requires protein synthesis and follows that of primary response genes.
- Expression requires synthesis of new proteins.

Secondary root : Lateral branch of the primary root.

Secondary transport : Active transport driven by the proton gradient established by the proton pump.

Secondary wall : Cell wall synthesized by nongrowing cells. Often multilayered and containing lignin, it differs in composition and structure from the primary wall. Forms during cell differentiation after cell expansion ceases.

Secretory vesicles : Derived from the golgi, carry the polysaccharides and glycoproteins to the plasma membrane, where the vesicles fuse with plasma membrane and empty their contents into the region of the cell wall.

Seed (s) :

- Mature ovule consisting of an embryonic plant together with stored food, all surrounded by a protective coat.
- Develops from the ovule after fertilization of the egg. Consisting of protective layers enclosing embryo of seed plants. May contain nutritive endosperm tissue separate from the embryo.

Seed additives : Additives are defined as all kinds of molecules or materials or mixes thereof which can have a positive influence on the seed germination or the growth and development of seedlings. Different types of additives are recognized such as micronutrients, plant growth regulators, activators of genes, activators of systemic resistance, plant growth promoters and miscellaneous types such as seaweed or plant extracts.

Seed analyst : A qualified person appointed under sec. 13 of the Act for carrying out seed quality analysis on submitted seed sample.

Seed bank : The seed population present in the soil or in the ecosystem acting as a source for perpetuation of species. Syn. Seed pools, soil seed bank.

Seed blower : An equipment used to separate light fractions in the seed sample prior to purity analysis.

Seed coat : The outer layer of the seed, derived from the integument of the ovule.

Seed coat dormancy : Dormancy imposed by impermeable seed coat.

Seed coating : Application of thin layer of inert material or chemical on the seed coat such that the basic shape of seed is not altered.

Seed colouring : Process of colouring seeds using organic or inorganic dyes so as to improve the appearance; to identify seed treatment varietal identification or brand imprint.

Seed count test : A test carried out to determine the number of seeds per kilo. This will help in assessing precise seed rate.

Seed density : It is the ratio of dry seed weight (100 seeds) in gram to the dry seed volume (100 seeds) in gram ml^{-1}

Seed dormancy : Viable seed fail to germinate due to internal factors.

Seed emptiness quotient : The ratio of empty seed to fully developed seeds

Seed enhancements : Refers to all those post-harvest treatments that improve germination or seedling growth, or facilitate the delivery of seeds and other materials required at the time of sowing.

Seed filling : Deposition of nutrients and storage materials in the storage organs of a developing seed.

Seed fingerprinting : A technique of assessing the presence of various seed proteins using poly acrylamide gel electrophoresis (PAGE). The banding obtained is unique for a variety as a fingerprint is unique to humans. This is a means of varietal identification.

Seed fortification : Soaking seed in nutrient solutions to improve germination and vigour.

Seed hardening : A presowing seed treatment to induce drought tolerance in seeds. The seeds are soaked in optimum quantity of chemical solution for 6-12hrs depending upon the seed and shade dried back to original moisture content. Syn. Pressowing hardening.

Seed infusion : A method of introducing seed treating chemicals into the seeds by dissolving the chemicals in organic solvents and soaking seeds in it; this does not alter the moisture content of seeds; it is also termed as dry permeation.

Seed leachate : Seed steep water that contains electrolytes leached out of seed.

Seed moisture : The quantity of water present in the seed either in liquid or vapour from.

Seed pelleting : Application of a thick coat of inert material that modifies the shape of seed and makes it amenable for further agronomic practices.

Seed plants : Plants in which the embryo is protected and nourished within a seed. The gymnosperms.

Seed polymorphism : Variation in the seed form or seed appearance due to various factors such as seed position; maturity levels etc. Syn. Heteroblasty.

Seed quality : Defined as the buildup of qualities considered being important for planting value of seeds.

Seed Replacement Rate (SRR) : A target of Government agency or seed company to replace the farmers seed with quality seeds of improved varieties or hybrids so as to augment agricultural productivity.

SRR = (A/B) x 100

Where A is Quantity of seed used;

B is Quantity of seed actually needed based on the total area of the crop.

Seed sampling : The process of taking representative samples for sending to seed testing lab.

Seed stimulation : A general term given to all those methods used for increasing the vigour of seeds.

Seed stock : Quantity of seed reserved for planting for the production of a new seed crop.

Seed technology : A broad based multi-disciplinary science beginning and ending with seed.

Seed test : A series of test such as purity, moisture content, germination and seed health conducted on a seed sample to determine its eligibility for certification.

Seed testing : Science of evaluating the planting quality of the seed.

Seed treating drum : A drum fixed to a horizontal shaft with a lid to add and remove seeds; seeds along with seed treatment chemicals are put into the drum and rotated manually to achive uniform coating.

Seed treatment : Process of treating seeds with chemicals for protecting seed or for invigourating it.

Seed trier : Long hollow tubes made of brass/iron that are used for obtaining samples.

Seed vernalisation : Application of low temperature treatment to seed before sowing.

Seed vigour :

- Is the sum total of all those properties in seed which upon planting, result in rapid and uniform production of healthy seedling under a wide range of environments including both favourable and stress conditions (AOSA, USA). All vigourous seeds are viable, but all viable seeds need not be vigorous.
- All those properties in seeds which upon planting result in rapid and uniform production of healthy seedlings under wide range of environments including both favourable and stress conditions AOSA (Association of official seed Analysts), USA (1975).

Seed vigour-bioassay : A direct vigour test that employs the effect of gases (aldehydes) released by aged seeds on sensitive seeds for rating vigour.

Seed volume : The volume of dry or soaked seed/s. This is taken by putting counted number of dry or soaked seeds in an indented cylinder with fixed amount of water in it. The increase in volume in milliliters is recorded for calculation of swelling capacity.

Seedcoat : Outermost tissues or "skin" of a seed; sometimes this coat is extremely hard and waterproof, preventing entrance of water to initiate germination unless it is broken, scratched, or eroded.

Seedling : Young plant grown from a seed.

Seedling growth rate : It is determined by dividing mean increase in length from previous measure by number of days the seedlings are placed in germinators.

Seedling phase : Recently germinated plants that still have their cotyledons attached.

Seedlot : A homogenous mass of seeds from which representative seed quantity can be drawn by random sampling.

Selectable markers : Markers used to select the transformed cells from among non-transformed ones. The transformed cells can easily be selected because only the cells containing the selectable marker gene can survive in the selection medium containing the toxic agent.

Selectively permeability (of a membrane) : Membrane property that allows diffusion of some molecules across the membrane to a different extent than other molecules.

Selectivity filter : The domain of a channel protein that determines its specificity of transport.

Self-assembly : Tendency for large molecules under appropriate conditions to aggregate spontaneously into organized structures.

Self-thinning : Reduction in plant density due to increased mortality as a result of competition.

Semiautonomous organelles : Organelles which have their own genomes and can divide independently of the cell but depend on the nucleus for the majority of their proteins. Examples - Chloroplast and mitochondria.

Seminal roots :

- Adventitious roots collapses after some time.
- In rice seedling in addition to the root system that develops from the radicle namely primary, secondary and tertiary roots, horizontal roots arising from mesocotyl numbering 2-6 known as seminal roots also arise. These are considered essential for a normal seedling. It serves to anchor the seedling when transplanted.
- Start of root development with the emergence of three to six primary root axes from the germinating seeds.

Semipermiable or selectively permeable membrane : Which allows only solvet particles to diffuse across not the solute.

Senescence :

- An active, genetically controlled, developmental process in which cellular structures and macromolecules are broken down and translocated away from the senescing organ (typically leaves) to actively growing regions that serve as nutrient sinks. Initiated by environmental cues, regulated by hormones.
- Normal, energy dependent developmental process that is controlled by plants own genetic program. Leaves are genetically programmed to die, and their senescence can be initiated by

environmental cues. Hydrolytic enzymes breakdown many cellular proteins, carbohydrates and nucleic acids. Substances obtained from breakdown are then transported back into the plant via phloem.

- Programmed series of metabolic events that involve metabolic breakdown of cellular constituents and transport of the breakdown products out of the senescing organ that ultimately dies.
- Shedding of plant parts in response to ageing results in degradation of fats, proteins, chlorophylls and nucleic acids in to simpler compounds like amino acids, fatty acids, glycerols, nucleotides, nucleosides. Soluble substances are exported out of leaves which become chlorotic, dries and falls down.

Senescence associated genes (SAGs) : Genes whose expression is induced during senescence.

Senescence down regulated genes (SDGs) : Genes whose expression decreases during senescence.

Senescent phase : Slow growth rate towards later growth periods.

Sense RNA : RNA capable of translation into a functional protein.

Sensible heat loss : Loss of heat from leaf surfaces to the air circulating around the leaf, under conditions in which leaf surface temperature is higher than that of the air.

Sepals : Green, leaf like structures that form the outermost part of a flower. In bud, they enclose and protect other flower parts.

Serotinous : State of cones on a tree that remain closed with release of seeds delayed or occurring gradually.

Serpentine soil : Soils that naturally contain high levels of various heavy metals and magnesium, but low concentrations of calcium, nitrogen, and phosphate.

Sesquiterpenes : Terpenes having fifteen carbons, three five- carbon isoprene units.

Sesquiterpene lactones : Bitter, antiherbivore, 15-carbon terpenes found in members of the composite family, such as sunflower and sage brush.

Sex determination : The process whereby unisexual flowers are produced by the early selective abortion of either the stamen or the pistil primordia. Genetically regulated, but also influenced by photoperiod and nutritional status. Mediated by GA.

Shade avoidance response :

- A response to shading; includes lengthening of the stem
- Plants that increase stem extension in response to shading.

Shallow dormancy : Dormancy that does not require after ripening and mere storage for few weeks will overcome it.

Shattering : Shedding of seeds from fruits before harvest.

Shattering loss : Loss of seed by shattering.

Shikimic acid pathway : Reactions that convert simple carbohydrate precursors to the aromatic amino acids-phenylalanine, tyrosine, and tryptophan.

Shoot apex : Consists of the shoot apical meristem plus the most recently formed leaf primordia (organs derived from the apical meristem)

Shoot apical meristem : Meristem at the tip of a shoot. Consists of tunica layers, central zone, peripheral zone, rib meristem.

Shoot axis : Principal line along which a shoot may grow, hypothetical central line of a shoot.

Shootmeristemless gene : Required for the formation of shoot promeristem.

Short day plants (SDP) :

- Flowers when day length is more than certain critical light period (less than 10 hours). Examples - Tobacco, Cocklebur, Cotton, Coffee, Soybean, Paddy, Maize, Sugurcane etc.
- A plant that flowers only in short days (qualitative SDP) or whose flowering is accelerated by short days (quantitative SDP).
- Plants whose flowering is induced by exposure to long nights.

Short - long day plants : Plants require shorter photoperiods for some days followed by longer photo periods. Examples - Winter rye, candytutf.

Short root and scarecrow genes : Required in the establishment of the radial tissue pattern in the root and hypocotyl during embryogenesis (Scheres *et al.* 1995; di Laurenzio *et al* 1996).

Short-distance transport : Transport over a distance of only two or three cell diameters. Involved in phloem loading. When sugars move from the mesophyll to the vicinity of the smallest veins of the source leaf, and in phloem unloading, when sugars move from the veins to the sink cells.

Shrivelled seeds : Immature seeds that show shriveling on account of drying.

Shrunken seed : Alternate wetting and drying of mature seed leading to shrinking.

Sieve cells : The relatively unspecialized sieve elements of gymnosperms. Contrast with sieve tube elements.

Sieve effect : The penetration of photosynthetically active light through several layers of cells due to the gaps between chloroplasts permitting the passage of light.

Sieve elements : Cells of phloem that conduct sugars and other organic materials throughout the plant.

Sieve element loading : The movement of sugars into the sieve elements and companion cells of source leaves, where they become more concentrated than in the mesophyll.

Sieve element-companion cell complex : A functional unit consisting of a sieve element and its companion cell.

Sieve elements unloading : The process by which imported sugars leave the sieve elements of sinks.

Sieve plates : Sieve areas found in angiosperm sieve- tube elements, they have larger pores (sieve-platepores) than other sieve areas and are generally found in end walls of sieve tube elements.

Sieve tube : Tube formed by the joining together of individual sieve tube elements at their end walls.

Sieve tube elements : The highly differentiated sieve elements typical of the angiosperms. Contrast with sieve cells.

Sigmoid curve : S shaped curve.

Signal transduction : A sequence of processes by which an extracellular signal (typically light, a hormone or neurotransmitter) interacts with a receptor at the cell surface, causing a change in the level of a second messenger and ultimately a change in cell functioning.

Signal-transduction pathway : Chain of events by which a chemical messenger (e.g., a phytohormone or other signaling molecule) or physical (e.g., radiation) signal is sensed and relayed into a chain of molecular events that lead to a response; it can operate at the cellular or wholeplant level, involving long-distance transport of the signal.

Simple lipids : Esters of monocarboxylic alephatic alcohals and yield only these compounds on hydrolysis.

Sin mutant : In peas, a mutant having abnormally high levels of GA_{20} in the seed due to the impairment of a hydroxylation step in GA deactivation. In barley, an ultra-tall mutant resulting from a recessive allele causing negation of a negative signal transduction factor.

Singlet oxygen ($^1O_2$) : An extremely reactive and damaging for of oxygen formed by reaction of excited chlorophyll with molecular oxygen. Causes damage to cellular components especially lipids.

Sink :

- Where the photosynthates are utilized.
- Any organ that imports photosynthates, including nonphotosynthetic organs and organs that do not produce enough photosynthetic products to support their own growth or storage needs, e.g. roots, tubers, developing fruits, and immature leaves. Contrast with source.
- Net importer of photoassimilates.
- Part of the plant that shows a net import of a compound (e.g., a root is a sink for carbohydrates and a leaf is a sink for inorganic nutrients).

Sink activity : Rate of uptake of photosynthates per unit weight of sink tissue

Sink size : The total weight of the sink.

Sink strength : The ability of a sink organ to mobilize assimilates towards itself. Depends on two factors : sink size and sink activity.

Sleeve trier with partitions : Similar to sleeve trier except that the inner tube is partitioned inside in between two holes; it is used to take samples from different depths when sampling from a heap or wagon.

Slop culture : Nutrient solutions are poured over solid medium.

Slotted screen : A perforated metal sheet or evenly spaced holes of desired size to grade seeds based on width.

Sodicity : Refers to the high concentration of Na^+ in the soil. Contrast with salinity.

Sodium chloride test : A screening test to measure the salt tolerance of varieties. Seeds are soaked in salt solution and then germinated to assess tolerance to salinity.

Sodium Hydroxide soak test-identification of pathogen : A technique for identification of wheat karnal bunt by soaking seed in 0.2% NaOH for one day and washing it; infected portion appears black in colour.

Sodium Hydroxide soak test identification of variety : A varietal identification test for wheat to distinguish between red and white wheat by soaking in 5% NaOH solution for one hour. Red varieties become orange brown while white varieties remain yellow.

Sodium-Dedocyl-Sulphate-Poly-Acrylamide Gel Electrophoresis (SDS-PAGE) : A technique of separating seed proteins by exposing them to anionic detergent i.e. SDS followed by electrophoresis by PAGE for more accuracy.

Soil analysis : Chemical determination of the nutrient content in a soil sample from the root zone.

Soil hydraulic conductivity : Measure of ease with which water moves through the soil and varies with type of soil and water content.

Soil texture : Particle size distribution in a soil, e.g., the relative proportions of sand, silt, and clay.

Sol : More viscous though liquid can be poured from one container to another. Example - Milk, latex.

Solar tracking :

- Movement of a leaf or flower that positions this organ at a more or less constant angle relative to the incident radiation throughout the entire day.
- The movement of leaf blades throughout the day so that the planar surface of the blade remains perpendicular to the sun's rays.

Solarization : Destruction or photooxidations of chlorophyhll under influence of high light intensities.

Solid medium culture : Nutrient solutions are added into the solid medium such as sand, crushed quartz or graval.

Solute : Dissolved substance.

Solute potential (or osmotic potential) Ψ_s : The effect of dissolved solutes on water potential.

Solution : Substances dissolved in any media.

Solvent : Dissolving media.

Somaclonal variation : Genetic variation generates while culturing the plant cells in vitro.

Somatic : Non reproductive.

Somatic embryogenesis :

- Haploid and diploid cells develop into differentiated plants through characteristic embryological stages.
- The process of asexual development of embryos similar in structure, function and biochemistry of true embryos (zygote) from plant tissue; somatic embryos are cultured in the lab.

Somatic embryos : Embryos arise from single cells located within clusters of meristematic cells either in callus mass or in suspension.

Somatic hybridization : A sexual cell fusion.

Sorption : When a molecular, ion or atom come in contact with some surface

Sound seed : A term used to define good, healthy seed without any damage.

Source :

- Where the photosynthates enter the plant or synthesized.
- Any exporting organ that is capable of producing photosynthetic products in excess of its own needs, e.g., a mature leaf or a storage organ. Contrast with sink.
- Net exporter of photoassimilates.
- Part of a plant that shows a net export of a compound (e.g., a leaf is a source for carbohydrates and a root is a source for inorganic nutrients).

Southern blot analysis : Method for detecting DNA fragments in an agarose gel by blotting on a nylon or nitrocellulose membrane followed by detection with a probe of complementary DNA or RNA sequences.

Soxlet apparatus : A glass apparatus used for accurate estimation of oil content of seeds using organic solvent like petroleum ether.

Spatial : Pertaining to space, the form and distribution of matter in space.

Specific heat :

- Energy required to raise the temperature of a unit mass of substance 1°C.
- Ratio of the heat capacity of a substance to the heat capacity of a reference substance, usually water. Heat capacity is the amount of heat needed to change the temperature of a unit mass by 1°C. The heat capacity of water is 1 calorie per gram per degree Celsius.

Specific leaf area (SLA) :

- Leaf area per unit leaf dry Mass
- Ratio of leaf area to leaf dry weight indicates relative proportion of assimilatory, conductive or mechanical tissues in leaves.

Watson (1952) $SLA = \frac{A}{WL}$ Area weight^{-1}

For two stages Gardner et al. (1985) : $SLA = \frac{\frac{A_1}{W_1} + \frac{A_2}{W_2}}{2}$

Specific leaf weight (SLW)/ Specific leaf mass :

- Leaf dry mass per unit leaf area (LMA).
- Ratio between leaf dry weight and leaf area

  Beadle (1982) $SLW = \frac{WL}{A}$ Weight Area^{-1}

  For two stages Gardner et al. (1985) : $SLW = \frac{\frac{WL_1}{A_1} + \frac{WL_2}{A_2}}{2}$

Specific root length (SRL) : Root length per unit root dry mass.

Spectrophotometer : Instrument that measures the amount of light absorbed at different wavelenghts by a substance.

Spectrophotometry : The technique used to measure the absorption of light by a sample.

Speed of germination :

- High speed of germination is an indication of a vigorous seed lot. It is estimated by germinated seeds counted daily and cumulative index is worked out.
- A mathematical expression similar as that of rate of germination but here only the normal seedlings are counted.

  $$\text{Speed of germination} = \frac{X_1(X_2 - X_1)}{Y_1(Y_2 - Y_1)} + \frac{X_n - X(n^{-1})}{Y_n - Y(n^{-1})}$$

  Where X is the germination percentage on n^{th} day and Y is the number of days from date of sowing.

Spike : A simple racemose type of inflorescence in which flowers are sessile.

Spikelet : A part of an inflorescence in grass family usually the unit of the flower including bracts and one or more flowers.

Spiral phyllotaxy : One leaf is produced at each node, with each subsequent leaf produced at a 137° angle to the previously formed leaf.

Split cotyledon : A type of injury caused to groundnut seed due to prolonged drying under sun leading to cracks in cotyledon that are visible through the testa.

Split seed coat bean : A phenomenon noticed when large seeded beans are accidentally crossed with small seeded types; the seeds of the small seeded type show splitting of seed coat caused by enlargement of the crossed embryo (cotyledon).

Spongy mesophyll :

- Loosely packed photosynthetic cells at the abaxial side of a dorsiventral leaf.
- Mesophyll cells of very irregular shape located below the palisade cells and surrounded by large air spaces.

Spongy parenchyma : Mesophyll cells which contain air spaces.

Spout : Outlets provided in processing machine to receive graded seeds.

S-protein : Type of protein present in self-incompatible lines and responsible for self-incompatibility.

spy mutant : In Arabidopsis, an ultra-tall constitutive response mutant resulting in the loss of function of a negative regulator.

$\Psi_s = -RTc_s$: Where R is the gas constant, T is the absolute temperature (in degrees Kelvin or K), and cs is the solute concentration of the solution expressed as osmolality.

Stamen : Male reproductive organ of the flower that produces pollen. Consists of stalk (filament) and an anther.

Staminate : Flower having only stamens but no carpels hence a typical male flower. Opp. Pistillate.

Starch : Polymer of glucose; storage compound in Plastids.

Starch - Sugar hypothesis : Opening and closing of stomata in response to accumulation of sugar in day and starch in night due to increase or decrease in pH of guard cells.

Starch sheath : A layer of cells that surrounds the vascular tissues of the shoot and coleoptile and is continuous with the root endodermis. Required for gravitropism in Arabidopsis shoots.

Static method : Procedure to measure the equilibrium moisture content of seed by placing them in desiccator containing salt solution or acid to get specific relative humibity.

Static pressure : Pressure exerted by a mass of seeds resting in a bin or drier that impedes the flow of forced air; used as a measure of penetration of air a drier for seed drying.

Statocytes : Specialized gravity-sensing plant cells that contain statoliths.

Statoliths :

- Amyloplasts that function as gravity sensors.
- Cellular inclusions such as amyloplasts that act as gravity sensors by having a high density relative to the cytosol and sedimenting to the botton of the cell.

Steady state :

- Condition in which influx and efflux of a given solute are equal. Therefore, the ion concentrations are constant with respect to time.
- The condition in which influx and efflux of a given solute between two compartments, separated by a membrane are equal. In the presence of active transport, steady state will differ from equilibrium.

Steam and dry sterilization : Sterilization of glassware and metallic instruments in dry heat for 3 hours at 160-180° C.

Stearoyl - ACP Desaturase (ACP-Acyl carrier protein) : Required for alternation in the ratio of saturated to unsaturated fatty acids.

Stele stem cells : Root stem cells immediately behind the quiescent center that give rise to the pericycle and vascular tissue.

Stem cells :

- Undifferentiated cells that retain the capacity for cell division indefinitely.
- Uncommitted, slowly dividing initial cells that produce all the cells in the meristem and thereby all the cells in the entire plant body (stem and root). Upon cell division, one daughter cell remains a stem cell, while the other is committed to a developmental pathway.

Stem crack in celery : During flowering due to Boron deficiency stem cracks are developed in some of the varieties.

Sterilization : Procedure for elimination of microorganisms.

Stigma :

- The part of pistil that receives pollen.
- The receptive surface for pollen atop the style.
- Upper end of ovary.

Stipule : A leafy appendage often paired and usually at the base of stalk.

STM gene : Arabidopsis SHOOT-MERISTEMLESS gene that suppresses cell differentiation, ensuring that shoot meristem cells remain undifferentiated. Function required for formation of shoot protomeristem.

Stoichiometry : The quantitative relationship between the amounts of reactants consumed and the products produced during a chemical reaction.

Stoma (plural stomata) :

- A micro-scopic pore in the leaf epidermis surrounded by a pair of guard cells and in some species, also including subsidiary cells. Stomata regulate the gas exchange (water and CO_2) of leaves by controlling the dimension of stomatal pore.
- Minute pores on leaf surface connected with gaseous exchange and transpiration.
- Structures in the leaf epidermis formed by specialized epidermal cells; mostly the term refers to the pores, as well as to the stomatal apparatus.

Stomatal transpiration : Loss of water in the vapour form from stomatal pore.

Stomatal aperture : The opening in the leaf epidermis through which gases enter and leave the interior space of the leaf. Changes in stomatal apertures are controlled by guard cells.

Stomatal complex : The guard cells, subsidiary cells, and stomatal pore, which together regulate leaf transpiration.

Stomatal conductance/resistance :

- Conductance/ resistance for transport of CO_2 or water vapour through the stomata.
- A measurement of the flux of water and carbon dioxide through the stomata, in and out of the leaf. The inverse of stomatal resistance.

Stomatal index : Determined as follows –

$$I = \frac{E}{E+S} \times 100$$

I = Stomatal index, S -= no. of stomata/unit area, E = No. of epidermal cells

Stomatal movements : Opening and closing of the stomata due to tugor changes in the guard cells.

Stomatal pore :

- An opening through the epidermis to the leaf interior. Surrounded by a pair of guard cells.
- Opening in the leaf epidermis between two guard cells of stomata.

Stomatal resistance (r_s) : A measurement of the limitation to the free diffusion of gases from and into the leaf posed by the stomatal pores. The inverse of stomatal conductance.

Storage : Build-up of a metabolically inactive pool of compounds that can subsequently serve to support growth or other physiological functions; see reserve formation.

Storage fungi : Species of fungi that harbour seeds during storage under limited moisture conditions and are mostly saprophytic.

Storage sinks : Where the imported issimilates are deposited in the form of storage compounds as sucrose, starch, protein or fatty acids (HO, 1988).

Strach-statolith hypothesis : Proposed mechanism for gravitropism involving sedimentation of statoliths in statocytes.

Straight fertilizer :

- fertilizer that contains only one of the nutrients among NPK Examples – Ammonium Nitrate, superphosphate.

- Chemical fertilizers that contain inorganic salts of only one of the three macronutrients : nitrogen, phosphorus, and potassium (e.g., superphosphate, ammonium nitrate, or muriate of potash, a source of potassium).

Straight-growth test : A bioassay based on the ability of auxin to stimulate the elongation of Avena coleoptile sections.

Strain : Affected physiological mechanism due to stress.

Strategy complex : Suite of traits allowing adaptation to a particular environment.

Stratification :

- A method of improving germination of those dormant seeds caused by immature embryo; such seeds are maintained in moist medium at high temperature (30°C), (Warm stratification) or low temperature (5-10°C) (cold stratification). Such exposure increases the endogenous production of hormones that promote embryo growth. Syn. Seed vernalization.
- Breaking of seed dormancy by exposure of moist seeds to low temperatures.
- In some plants, a cold temperature requirement for seed germination. The term is derived from the former practice of breaking dormancy by allowing seeds to overwinter in small mounds of alternating layers of seeds and soil.
- Incubating seeds/plant material to low temperature i.e. O -5°c to break dormancy or improve performance.

Stress :

- Any change in environmental conditions that might reduce or adversely change a plants growth and development (its normal function) Levitt (1974). Or an external factor that exerts a disadvantageous influence on the plant.
- Disadvantageous influences exerted on a plant by external abiotic or biotic factor(s), such as infection, or heat, water and anoxia. Measured in relation to plant survival, crop yield, biomass accumulation, or CO_2 uptake.
- Environmental factor that reduces plant Performance.

Stress protein : Protein that is produced only or in greater quantities upon exposure to stress.

Stress relaxation : Selective loosening of bonds between primary cell wall polymers, allowing the polymers to slip by each other, simultaneously increasing the wall surface area and reducing the physical stress in the wall.

Stress response : The immediate detrimental effect of stress on a plant process causing reduced plant performance.

Stress test : Test in which seeds are intentionally subjected to the action of adverse external forces, usually unfavourable germination conditions.

Stress tolerance (stress resistance) : A plant's ability to cope with an unfavourable environment.

Stroma :

- Fluid compartment surrounding the thylakoids.
- Matrix within the chloroplast containing Calvin-cycle enzymes and in which the thylakoid membrane system is suspended.

Stroma lamellae :

- Exposed membranes in which stakking is absent.
- Unsacked membranes that connects adjacent grana.
- Unstacked thylakoid membranes within the chloroplast.

Stroma reactions : The carbon fixation and reduction reactions of photosynthesis that take place in the stroma of the chloroplast.

Strong electrolytes : 100% molecules dissociate in to ions when dissolved. Example - NaCl etc.

Strophiole (¼caruncle) : An outgrowth of a seed coat, near the hilum; preformed weak site in the seed coat that allows entry of water when sufficiently weathered.

Style :

- A stalk-like extension of the carpel.
- Neck of ovary.

S-type channels : A type of gated channel for anions that remains open for the duration of the stimulus.

Subculture : Transplanting a cell, tissue or organ etc. from one nutrient solution to the another.

Suberin :

- Found in cork cells and walls of epidermal cells.
- Polymer, structure is poorly understood, formed from hydroxy or epoxy fatty acids joined by ester linkages. Differs from cutin in that it has dicarboxylic acids, more long chain compounds and a significant proportion of phenolic compounds as part of its structure.
- A wax-like lipid polymer similar to cutin that acts as a barrier to water and solute movement through the Casparian strips of the endodermis, the outer cell walls of underground orgens, cork cells of the periderm, and sites of leaf abscission or wounding.
- Polymer containing long-chain acids, hydroxy acids, alcohols, dicarboxylic acid, and phenols; the exact structure is not fully understood; cell-wall component in many locations (e.g., Casparian strip, corky periderm).

Subirrigation system : Nutrient solution is forced up from the bottom of the container by using pumping apparatus.

Subjective day : When an organism is placed in total darkness, the phase of the rhythm that coincides with the light period of a preceding light / dark cycle.

Subjective night : When an organism is placed in total darkness, the phase of the rhythm that coincides with the dark period of a preceding light/dark cycle.

Submitted sample : The part of composite sample mixed and divided or reduced to the required submitted sample size and sent to the seed testing laboratory for testing.

Sub-optimal temperature : Temperature below the normal (optimal) germination temperature.

Sub-sampling : The method of obtaining working samples by mixing and dividing the submitted samples.

Subsidiary or accessory cells :

- Epidermal cell type around many stomata, located distally and laterally to a guard cell.
- Specialized epidermal cells that flank the guard cells and work with the guard cells in the control of stomatal apertures.
- Pair of differentiated epidermal cells which flanks guard cells.

Substomatal cavity : The air space within the leaf to which the stomatal pore opens and which is bounded by mesophyll cells.

Substrate : Substratum, a substance acted upon,

Substratum : The medium used for germination.

Subunit : A polypeptide that is part of a protein complex.

SUC2 protein : A sucrose-H^+ symporter found in the plasma membrane of companion cells.

Succession : Directional change in plant species composition resulting from biotically driven changes in resource supply.

Succulence : Thick fleshy state of herbaceous tissues due to high water content; it is quantified as the volume of water in the leaf at a relative water content of 100% divided by the leaf area.

Succulent plants :

- Plant with tissue of high degree of Succulence
- Stomata open during night and close during day due to accumulation and removal of malic acid which acts as strong solute.

Sucrose-H^+ symporter : An active carrier that couples the energy dissipated by a proton, diffusing back into the cell to the uptake of a sucrose molecule. Mediates the transport of sucrose from the apoplast into the sieve element-companion cell complex in source leaves of apoplastically loading species.

Sugar sensing : The perception of internal sugar concentrations that is at the start of a signaltransduction pathway.

Sui generic : (sul=self; generic=generating) i.e. System of it's own kind. The exclusive right of breeder over his variety to produce and distribute the seeds.

Sulpholipids : Lipids contain one or more molecules of sulphur.

Summer annual : Species whose seeds germinate after winter and completes its life cycle before the start of the next winter.

Sunflecks :

- Short period of high irradiance that interrupts the background of low diffuse radiation in and under leaf canopies caused by direct sunlight that penetrates small holes in the canopy.
- Patches of sunlight that pass through openings in a forest canopy to the forest floor. Major source of incident radiation for plants growing under the forest canopy.

Supercooling : Refers to the noncrystalline state of water at sub-zero temperatures.

Superoxide ($O_{.2}^{-}$) : A reduced form of oxygen that is damaging to biological membranes.

Superoxide dismutase (SOD) : This enzyme converts superoxide radicals to hydrogen peroxide.

Supersaturated solution : Concentration of solute is greater than saturated solution.

Supply function : Equation describing CO_2 diffusion from the atmosphere into the leaf, supplying substrate for photosynthesis.

Supra optimal temperature : Temperature above the normal (optimal) germination temperature.

Surface sterilization : Use of chemicals like sodium hypochloride, calcium hypochloride, H_2O_2, $AgNo_3$, Bromine water etc. for sterilization of explants.

Surface tension : A force exerted by water molecules at the air- water interface, resulting from the cohesion and adhesion properties of water molecules. This force minimizes the surface area of the air-water interface.

Suspension : Clay + water type system with solid dispersed phase and liquid continuous phase. Example - Ink.

Suspensor :

- A group or chain of cells from the zygote, which pushes the developing pro embryo towards the center of the ovule and also supplies nutrient to the developing zygote.
- Filament of six to nine cells that attaches the embryo to the vascular system of plant. Only one of the basal cells derivative contributes to the embryo called hypophysis and it forms columella or central part of the root cap and an essential part of root apical meristem known as quiescent center.
- In seed plant embryo-genesis, the structure that develops from the basal cell following the first division of the zygote. Supports, but is not part of, the embryo that develops from the apical cell.
- Lower end of embryo during embryo developmental stage consists of row of cells. Main function is to push the developing embryo in to endosperm, disintegrates very soon.

SUTI protein : One of several sucrose-H^+ symporters found in the plasma membrane of sieve elements.

Swelling capacity : It is the difference in seed volume of soaked and dry seeds.

Swelling index : It is the ratio of swelling capacity (100 seeds) to the dry seed weight (100 seeds) in grams.

Swollen seeds : Seeds which have imbibed water and appear healthy but remain ungerminated at the end of test period. Syn. Fresh ungerminated seed.

Symbionts : Either one of the two organisms associated in a symbiotic relationship that may or may not be mutually beneficial.

Symbioses, singular symbiosis :

- Alliance between roots and microorganisms for their mutual benefit.
- Intimate association between two organisms of different species (in this text, the term is used when both symbionts derive a long-term selective advantage).

- The close association of two organisms in a relationship that may or may not be mutually beneficial. Often applied to beneficial (mutualistic) relationships.

Symbiosome : Membrane-surrounded space containing one or more rhizobia in an infected cell of a root nodule in a legume.

Symbiotic N fixation : Nitrogen fixation takes place when the organisms live together benefiting each other. Examples nodule forming rhizobium in roots of leguminous plants and certain fungi belonging to actinomycetes.

Symplast :

- It includes all living parts of plant cells viz. plasmodesmata and elements within cytoplasmic membrane.
- Space comprising all the cells of a plant's tissues connected by plasmodesmata and surrounded by a plasma membrane.
- The continuous system of cell protoplasts interconnected by plasmodesmata.

Symplast pathway : The route by which xylem and phloem sap travel from one cell to the next via the plasmodesmata.

Symplastic phloem loading : Occurs in plants in which photosynthates moves from the cytoplasm of the mesophyll cells of the leaves, via plasmodesmata, to intermediary cells; after chemical transformation into oligosaccharides, these move, again via plasmodesmata, to the sieve tubes.

Symplastic transport : Intercellular transport of solutes through plasmodesmata.

Symport :

- A type of secondary active transport in which two substances are moved in the same direction across a membrane.
- Co-transport of one compound in one direction coupled to transport of another compound (mostly H+) in the same (uniport) or opposite (antiport) direction.
- Two substances moving in the same direction through the membrance. Protein involved is called symporter.

Symporter :

- A protein carrier involved in symport.
- An integral membrane protein involved in movement of two or more different molecules or ions across a phospholipid membrane against the concentration gradient.

Synergids : Group of 3 nuclei located at the lower end of embryo sac.

Syntaxins : Proteins that integrate into membrances, permitting the membrances to fuse.

Synthetic auxin : A substance with auxin activity that is not produced by plants, for example, 2,4-D and dicamba. Often used as herbicides because they are not degraded by the plant as quickly as IAA.

Synthetic variety : A variety obtained by mating in all possible combinations of a number of lines that combine well with each other.

Systemic acquired resistance (SAR) :

- When a plant survives infection of a pathogen at one site, it often develops increased resistance to the subsequent attacks at sites throughout the plant and enjoys protection against a wide range of pathogen species (Ryals *et al.*1996).
- The increased resistance throughout a plant to a range of pathogens following the infection of a pathogen at one site.
- Resistance that is induced by a herbivore or a microorganism at a location that differs from the plant part that has been primarily affected; the organisms that induce the resistance may be parasitic or have a growthpromoting

Systemin : A polypeptide plant hormone that signals production of some defenses not continuously present in plants; initiates the biosynthesis of jasmonic acid.

❒❒❒

T

T1 plasmid : A circular DNA molecule from the Agrobacterium tumefaciens.

T50 : Mean germination time; the time taken (in days) for 50% of the seeds to germinate.

Tannin :

- Class of protein-precipitating polymeric phenolic secondary plant compound; typically a quantitative defense compound.
- Plant phenolic polymers with defensive properties. Tannins bind the collagen proteins of animal hides increasing their resistance to heat, water and microbes. Condensed tannins formed by polymerization of flavonoid units, frequent constituents of woody plants. hydrolyzable tannins are heterogenous polymers containing phenolic acids, especially gallic acid and simple sugars.
- Plant phenolic polymers that often bind proteins and function as defenses against microorganisms, insects, and many mammals, such as cattle, deer, and apes.

Tap root :

- Main single root axis in dicots directly originated from the radicle.
- Roots directly developed from radicle.
- The main single root axis from which lateral roots develop.

TATA BOX : A short sequence that determines the transcription start site named for its most abundant nucleotides, a component of promoters for genes in eukaryotic organisms that encode proteins.

T-DNA : A small portion of the Ti plasmid that is incorporated into the nuclear DNA of the host plant cell carries genes necessary for the biosynthesis of trans-zeatin and auxin, and opines. Because their promoters are plant eukaryotic promoters, none of these T-DNA genes are expressed in the bacterium, but are transcribed after they are inserted into the plant chromosomes.

Tegmen :

- Inner seed membrane
- The inner seed coat formed by the modification of inner integument of the ovary.

Temperate recalcitrant : Seeds that are recalcitrant and sensitive to desiccation but are tolerant to low temperature.

Temperature compensation :

- Oscillator which is unaffected by temperature and enables the clock to function normally under a wide variety of seasonal and climatic conditions.
- The feature that enables the clock to keep time at

Temperature compensation point : The temperature at which the amount of CO_2 fixed by photosynthesis, equals the amount of CO_2 released by respiration, in a given time interval.

Tempering : Is the time interval provided in between two successive drying process in order to stabilize the moisture content of seeds by diffusion, so that drying process in the seed will be uniform in all its parts.

Temporary or transient wilting : Despite of sufficient moisture in soil plants may exhibit sign of willing generally in a afternoon of a hot summer day when transpiration exceeds absorption and recover again at night. Associated with the diurnal fluctuation of water content.

Tensegrity is an architectural term : A contraction of tensional integrity, also refers to structural integrity created by interactive tension between the structural components.

Tensegrity model : New model for gravitropism proposed by Andrew staehelin and colleagues (Yoder *et al.* 2001).

Tensile strength : The ability to resist a pulling force. Water has a high tensile strength.

Tension : Often used to refer to the partial pressure of a gas.

Teratomas : Tumours that contain partially developed structures. Mutated tmr locus of T-DNA produced teratomas with abnormal proliferation of roots.

Terminal bud :

- Bud growing at the end of a branch or stem.
- Shoot apex enveloped by several or smaller number of undifferentiated leaves.

Terminator seeds : A common name given to all those transgenic seeds that will not produce germinable seeds after the first generation (or) Terminator is the name critics have given to a genetic engineering technique that renders second generation seeds sterile, preventing farmers from saving seed from their harvest and forcing them to buy new seed each year.

Terpenes (terpenoids or isoprenoids) :

- A large class of plant compounds formed from five-carbon isoprene units, many of which are secondary metabolites with antiherbivore activity
- Substances insoluble in water, biosynthesized from acetyl COA or glucolytic intermediates. Derived from the union of five carbon elements that have the branched carbon skelton of isopentane. Structural elements of terpenes are also, called as isoprene units. All terpenes are occasionally referred to as isoprenoids.

- Class of secondary plant compounds containing C and H, produced from the precursor mevalonic acid.
- Class of plant lipids that includes carotenoids and sterols.

Test weight : Average weight of eight replicates of hundred seeds showing a co-efficient variation less than 6 for chaffy seeds and 4 for non-chaffy seeds.

Testa :

- Outer seed membrane
- The seed coat formed by the modification of outer integument of the ovary.

Tetraterpenes : Terpenes having forty carbons, eight five-carbon isoprene units. The latter compound serves as building blocks for anabolic processes while the excess is excreted as acetate.

Tetrazolium salt : Common term used to denote 2,3,5 Triphenyl Tetrazolium chloride salt used for staining viable seeds.

Tetrazolium test :

- A test to assess the viability of seeds. Pre-imbibed seeds are soaked in tetrazolium salt solutions (0.5-1%) for 3-6hrs at 30-35^{0}C and assessed for staining pattern. Viable embryos are stained scarlet red in colour. Un-viable embryos do not stain. **Syn.** Quick test, Tz test, Topographical tetrazolium test.
- Quick test to determine seed viability.

Texture : The relative proportion of various forms of starch grains in an endosperm.

Thawing : The process of bringing cryogenically (frozen) stored seeds to room temperature.

The torpedo stage embryo : Stage forms as a result of cell alongation throughout the embryo axis and further development of the cotyledons.

Thermal hysteresis : Phenomenon in which the liquid to solid transition is promoted at a lower temperature than the solid to liquid transition.

Thermal hysteresis proteins (THPs) : Proteins that have the property of transition from liquid to solid is promoted at a lower temperature then is transition from solid to liquid.

Thermal induction : Changes in the growth of plants caused by exposure to given temperature leading to flowering.

Thermal insulation : The methods of preventing movement of hot air from outside into the seed storage room by lining wall with non-conductive material.

Thermodormancy : Dormancy induced by high temperature. Ex - Tobacco, lettuce.

Thermogenic respiration : Respiration that increases the temperature of an organ, such a the flowers of Arum lilies.

Thermogradient : Rate of change in temperature with increase in altitude or change in distance from a heat source or cooling source.

Thermohygrometer : An instrument that automatically records the temperature and relative humidity of the environment continuously for a week.

Thermoinhibitor : Inhibition of germination caused by high temperature. E.g. Lettuce.

Thermolabile : A substance which disintegrates upon heating.

Thermoperiodicity : Response of plants to varying diurnal and nocturnal temperature.

Thermostat : An instrument to maintain the temperature at a set level in any equipment.

Thermotolerance : Tolerance induction to otherwise lethal temperature through periodic brief exposure to subleathal heat stresses.

Thief trier : A small pointed hollow with opening on one hole to collect sample from bags.

Thigmomorphogenesis : Altered growth of plant organs in response to a physical force (touch, wind, vibrations, rain, turbulent water flow).

Thigmotropism :

- Plant growth in response to touch, enabling roots to grow around rocks and shoots of climbing plants to warp around structures for support.
- Growth with respect to touch, enables roots to grow around rocks and is responsible for the ability of the shoots of climbing plants to wrap around other structures for support.

Thin seed coat with mucilaginous layer : A classification of non-endospermic seed with embryos that are bent or spatulate and endosperm is thin; the mucilage layer prevents the diffusion of oxygen and prevents germination.

Thio urea (NH_2-CS-NH_2) : A chemical that promotes germination and substitutes for light and temperature requirement.

Thiosulphate (R-S^-) : An intermediate in the reduction of sulphate. Formed from thiosulphonate. Reacts with O-acetylserine to form cysteine.

Thiosulphonate (R-SO^-_3) : An enzyme-bound intermediate in the reduction of sulphate. Formed from APS.

Thousand seed weight : Weight of one thousand seeds expressed in grams.

Threshold : The magnitude of a stimulus that needs to be exceeded to elicit a response.

Thylakoid :

- Photosynthetic membrane suspended in the stroma in chloroplasts; it encloses a lumen and contains the photosynthetic pigments, electron-transport chain components and ATPsynthase.
- The specialized, internal, chlorophyll containing membranes of the chloroplast where light absorption and the chemical reactions of photosynthesis take place.

Thylakoid reactions : The chemical reactions of photosynthesis that accur in the specialized internal membranes of the chloroplast (called thylakoids). Include photosynthetic electron transport and ATP synthesis.

TIBA(2,3,5- triiodobenzoic acid) : A competitive inhibitor of polar auxin transport.

Tip growth : Localized growth at the tip of a plant cell caused by localized secretion of new wall polymers. Occurs in pollen tubes, root hairs, some sclerenchyma fibers, and cotton fibers, as well as moss protonema and fungal hyphae.

tir3 mutant : Arabidopsis (transport inhibitor response 3) : mutant that exhibits reduced polar auxin transport and NPA binding. tir3 contains a mutation in the BIG calossion-like protein, which is also defective in the doc1 mutant.

Tissue : Organized cell masses.

Tissue culture : Growth of isolated plant cells, tissues, or organs in the laboratory on an artificial medium consisting of various essential minerals, vitamins, hormones, and a carbon source.

Tissue tension : Result of differences in turgor and/ or cell-wall elasticity between different cells in a tissue or organ; the tension is relaxed when the organ is cut, resulting in deformation; tissue tension plays an important role in the closing mechanism of the carnivorous Venus fly trap (Dionaea).

Tissue-mass : Density dry mass per unit volume of a tissue.

tmr loci : Mutations in these sites of T-DNA resulted in an abnormal proliferation of shoots. "Shooty", mutations.

tmr locus : Mutations at this locus of T-DNA produced teratomas with abnormal proliferation of roots. "Rooty" mutations.

Tocopherol : Refers to a group of compounds known as vitamin E. A naturally occurring anti-oxidant that delays ageing of cells.

Tolerance :

- A permitted deviation (plus or minus) from the mean; in seed testing it refers to the permissible deviation among replications of the test results; results showing deviation beyond the tolerance limit are rejected.

- Endurance of unfavorable environmental Conditions.

- Plant simply tolerates or endures adverse environmental conditions through phenotypic modifications and changes in anatomical and histological structures with biochemical constituents. Or plant fitness to cope with an unfavourable environment.

Tonoplast : Membrane that surrounds vacuole.

Top of paper (TP method) : A method of testing seeds for germination where seeds are placed on a wet germination paper placed in any suitable container; this method is recommended for germination.

Top of sand (TS method) : A method of germinating seeds by placing them on sterilized sand taken in a suitable container.

Topping : A method advocated in jute seed production that involves cutting of the plant 45 days after sowing, to increase pod number and seed yield.

Torpedo stage embryo : The third stage of embryogenesis. The structure produced by elongation of the axis of the heart stage embryo and further development of the cotyledons.

Torus : A central thickening found in the pit membranes of tracheids of most gymnosperms.

Totipotency :

- Potential of cells or tissues to form all types of cells or to regenerate a plant.
- In differentiated cells, retention of full genetic capacity for the development into a complete plant.
- Potentiality of a nonembryonic cell to differentiate into an embryonic cell and then to develop into a complete new plant if environment is suitable.
- The ability to regenerate a plant from single somatic cell.

Towel papers : Specially designed brown rough and thick paper used for seed germination.

Toxic zone : With increase in nutrient concentration beyond the adequate zone, growth or yield declines because of toxicity.

Toxicity effects :

- Effects of injurious concentrations of ions particularly Na+, Cr or So_4^{-2} accumulates in cells.
- Injuries caused by high concentrations of ions (particularly Na^+, $C1^-$, or SO_4^{2-}) that accumulate in cells at low water potentials.

Tracheary elements : Water transporting cells of the xylem.

Tracheids :

- Cell type in the xylem.
- Spindle-shaped, water-conducting cells with tapered ends and pitted walls without perforations found in the xylem of both angiosperms and gymnosperms.

Trade-off : Balancing of investment in mutually exclusive traits (e.g., protective structures vs. photosynthetic machinery in leaves).

Trans-acting factors : Regulatory sequences which bind specific proteins that modulate the activity of the general transcription factors that assamble around the transcription start site with RNA polymerase II.

Transamination : Reversible reactions catalyzed by transaminases in which nitrogen as an amino group is transferred from an á-amino acid to an á-keto acid.

Transcellular current : A positive current, due largely to calcium ions, that flows into the cell on one side and loops back through the external medium. Establishes polarity prior to morphogenesis in Fucus zygotes, root hairs and germinating pollen grains.

Transcription : Synthesis of an RNA polymer bearing a base sequence that is complementary to a specific gene.

Transcription factors :

- Proteins that have an affinity for DNA. They are able to turn expression of genes on or off by binding to DNA sequences.
- Proteins that interact with DNA promoter elements to modulate gene expression (RNA transcription).

trans-cyclooctene : A strong competitive inhibitor or ethylene binding.

Transfer cells :

- Cell involved in transport that has a proliferation of the plasma membrane causing surface enlargement (e.g., in the phloem of plants using the apoplastic phloem-loading pathway, in the epidermis of aquatic plants using bicarbonate).
- A type of companion cell similar to an ordinary companion cell, but with finger like projections of the cell wall that greatly increase the surface area of the plasma membrane and increase the capacity for solute transport across the membrane from the apoplast.

Transgenes : Genes obtained from some unrelated organisms.

Transgenic plants :

- A plant expressing a foreign gene introduced by genetic engineering techniques.
- Plants containing transgenes.
- That contain one or more genes transferred from different kind of organism.

Transit peptide : An N-terminal amino acid sequence that facilitates the passage of a precursor protein through both the outer and the inner membranes of an organelle such as the chloroplast. The transit peptide is then clipped off.

Translation : Process whereby a specific protein is synthesized from amino acids according to the information encoded by the MRNA.

Translocation :

- The movement of photosynthates from sources to sinks in the phloem.
- Transport of solutes through the Phloem.

Translocator : A membrane based protein that functions as a carrier in the transport of one or more substances across the membrane.

Transmembrane pathway : The route followed by the xylem sap that sequentially passes from cell to cell, crossing the cell plasma membrane both on entering and exiting the cell. Transport across the tonoplast may also be involved.

Transmittance : Fraction of radiation incident on a body that passes through the body; mostly used with reference to leaves.

Transpiration :

- Loss of water from aerial parts of plant. Vital phenomenon regulated by cells.
- The evaporation of water from the surface of leaves and stems.
- Water loss from leaves or whole plants due to evaporation from within a leaf or other plant parts.

Transpiration ratio :

- The ratio of water loss to photosynthetic carbon gain. Measures the effectiveness of plants in moderating water loss while allowing sufficient CO_2 uptake for photosynthesis.
- Ratio of amount of water transpired by the plant to the amount of CO_2 assimilated by photosynthesis. C_3 plants have transpiration ratio 500.

$$TR = \frac{500\,\text{Water}}{1\,CO_2}$$

Transport : Molecular or ionic movement from one location to another; may involve crossing a diffusion barrier such as one or more membranes.

Transport proteins : Membranes containing proteins that facilitates the transport of selected ions and polar molecules.

Trans-zeatin : The principal free cytokinin, chemically similar to kinetin. Exogenously applied in presence of auxin, induces cell division in callus cells and promotes bud or root formation from callus cultures. Endogenous trans-zeatin delays senescence of leaves and promotes expansion of dicot cotyledons.

Treated seeds : Seeds applied with either chemicals, pesticides, colour or inoculum, which has not resulted in significant change in the weight or appearance, so that the original seed is easily identifiable.

Triacylglycerols (triglycerides) : Three fatty acyl groups in ester linkage to three hydroxyl groups of glycerol. Fats and oils.

Tricarboxylic acid : Three carboxyl groups with six carbon acids. Examples - Citric acid, isocitric acid etc.

Tricarboxylic acid cycle (TCA cycle) : Conversion of malate or pyruvate to CO2 within the mitochondria.

Trichome : Epidermal hair on a leaf or stem.

Triglyceride : Three hydroxyl groups of glycerol combines with fatty acids.

Triphenyl formazan : A red coloured; non-diffusable; water insoluble end product deposited in living tissues when 2,3,5 Triphenyl tetrazolium chloride is reduced in the presence of dehydrogenase enzymes; it can be extracted using methyl cellosolve. Syn. Formazan.

Triple response : Consists of inhibition and swelling of hypocotyl, inhibition of root elonation and exaggeration of the apical hook in Arabidopsis.

Triplet state or Metastable state : Two electrons in two different orbits spinning in the same direction in chlorophyll molecule.

Triploid : Chromosome no. (3 n)

Triterpenes : Terpenes having thirty carbons, six five-carbon isoprene units.

Tropism : Oriented plant growth in response to a perceived directional stimulus from light, gravity or touch.

True leaves : Leaves produced by the seedling other than the cotyledonary leaves.

True potato seeds (TPS) : Zygotic seed of potato formed by fertilization of male and female gametes; it is disease free; 150gm is sufficient for one hectare and is true -to-type.

True solution : Transparent, higher boiling, lower freezing point and water potential.

Truthful label seeds : Seeds of both notified and unnotified varieties that were not brought under certification but tested for germination and purity by the producer; provided with opal green label for authenticity.

Trypsin : Protein-hydrolyzing enzyme (in animals).

Tryptamine (TAM) : An intermediate in one of three biosynthetic pathways for the synthesis of IAA in plants. Formed by decarboxylation of the amino acid tryptophan.

Tube cell : Cell of pollen grain that collapses at the time of pollination.

Tuber hardening inspection : An inspection stage in potato seed crop where the tubers formed start to accumulate starch and harden.

Tumour : A mass of rapidly dividing, undifferentiated and disorganized cells.

Tunica layers : The outer cell layers of the shoot apical meristem. Outer most tunica layer generates the shoot epidermis.

Turbidity test : A simple vigour test to estimate the vigour. Seeds are soaked in twice the volume of water for 12 hours and observed for turbidity of the seed steep water. If the water is too turbid, then it inferred that the seeds are old, if transparent, then the seeds are vigorous.

Turgid : State of cell after water entry.

Turgor :

- Positive hydrostatic pressure in live plant Cells.
- The firmness of a cell resulting from its hydrostatic or turgor pressure.

Turgor pressure (or hydrostatic pressure) :

- Force per unit area in a liquid. In a plant cell, the turgor pressure pushes the plasma membrane against the rigid cell wall and provides a force for cell expansion.
- Pressure exerted by protoplast over cell wall.

Twisted seedling : A type of abnormal seedling with the shoot portion twisted around its axis.

Tyndall cones : Cones formed due to light scattering by colloidal solutions.

Type - B ARR genes : Genes which contain a transcription factor domain in addition to the receiver domain.

Type A - ARR (Arabidopsis response regulator) genes : Genes which are made up of solely of a receiver domain.

Tz test : Shortened form of Tetrazolium test.

□□□

Ubiquinone : A mobile electron carrier of the mitochondrial electron transport chain. Chemically and functionally similar to plastoquinone in the photosynthetic electron transport chain.

Ubiquitin : A small polypeptide that is covalently attached to proteins by the enzyme ubiquitin ligase using energy from ATP and that serves as a recognition site for a large proteolytic complex, the proteasome.

Ubiquitination : The tagging of a protein with the small protein ubiquitin, thus targeting the protein for destruction by the proteasome.

Ultra violet test : A test to determine the purity of annual rye grass whose roots emit fluorescence on exposure to UV light and not so with perennial rye grass roots.

Ultraviolet rays : Wavelengths shorter than 390 nm are called ultraviolet rays.

Ultraviolet sterilization : Radiation treatment for sterilizing plastic wares etc.

Uncoupler :

- A chemical compound that increases the proton permeability of membranes and thus uncouples the formation of the proton gradient from ATP synthesis.
- Chemical compound that enhances the membrane conductance for protons and so uncouples electron transport from phosphorylation.

Uncoupling : A process by which coupled reactions are separated in such a way that the free energy released by one reaction is not available to drive the other reaction.

Uncoupling protein : A protein that increases the proton permeability of the inner mitochondrial membrane and thereby decreases energy conservation.

Uniformity of germination : Number of days taken to achieve a germination between 15-85% of the germinated seeds.

Uniformity of seedlot : Refers to absence of any apparent variation in a seedlot.

Unit leaf rate (ULR) : Synonym for net assimilation rate (NAR).

Unit of certification : The extent of seed crop planted by an individual to be inspected during an inspection as a single unit or the extent of seed crop in single sowing report.

Unsaturated fatty acids : Double bond between carbon atmos, possess low melting point Examples linoic acid ($C_{17}H_{29}COOH$) and lenolenic acid ($C_{17}H_{29}COOH$).

Up-regulation : Increase in the normal rate of a process, sometimes involving increased transcription of genes encoding enzymes involved in that process.

Ureide exporters : Legumes of tropical origin that convert toxic ammonia to ureides such as allantoin, allantoic acid and citrulline for transport to the shoot via the xylem. Contrast with amide exporters.

Utilization sinks : Where most of the imported assimilates is used for growth. Example - meristems, roots etc.

UV-B photoreceptor : One or more unidentified compounds (technically not pigments) that absorb ultravoilet radiation between about 280 and 320 nm.

□□□

Vacuolar H⁺-ATPase : A large, multi-subunit enzyme complex, related to F_0F_1-ATPases, present in endomembranes (tonoplast, Golgi). Acidifies the vacuole and provides the proton motive force for the secondary transport of a variety of solutes into the lumen.

Vacuole :

- Large, water filled structure in cell that can occupy 80-90% of the total volume of the cell.
- Membrane-bound cell compartment filled with water and solutes; among others used for storage of sugars, nutrients, and secondary metabolites.

Vacuum seed counter : A device to count seeds that uses vacuum to adhere seed to the counting head and dropping them on to the germination medium upon release of the vacuum.

Vacuum storage :

- Storage in a sealed container from which all or nearly all air was evacuated before sealing, storage in a (partial) vacuum.
- Storage of seeds in a sealed container from which air is removed.

Validity : The period upto which the tag issued is valid; normally nine months from the date of test for all seeds.

Van't Hoff equation : Relates the solute potential, $Ø_s$, to the solute concentration.

Vanadate : Inhibitor of the H^+-ATPase. The H^+-ATPase is phosphorylated as part of the catalytic cycle that hydrolyze ATP. Because of this phosphorylation step, the plasma membrane ATPases are strongly inhibited by orthovanadate (HVO_4^{2}), a phosphate (HPO_4^{2-}) analog that competes with phosphate from ATP for the aspartic acid phosphorylation site on the enzyme.

Vapour pressure deficit (VPD) : Difference in actual vapour pressure and the vapour pressure in air of the same temperature and pressure that is saturated with water vapour.

Vapour pressure difference (Dw) : Difference in vapour pressure between the intercellular spaces and the atmosphere.

Vascular bundles :

- Group of cells, viz. Xylem, phloem and cambium generally associated with transportation of water and food materials and secondary growth.
- Strands of primary phloem and xylem, separated by the vascular cambium and often surrounded by a bundle sheath, found in shoots, but continuous with the vascular cylinder of the root.

Vascular cambium :

- A lateral meristem consisting of fusiform and ray stem cells, giving rise to secondary xylem and phloem elements, as well as ray parenchyma.
- Secondary meristem that differentiates alongwith the primary vascular tissue from the procambium within the vascular cylinder.

Vascular tissues : Plant tissues specialized for the transport of water (xylem) and photosynthetic products (phloem).

V-ATPase : A vacuolar H^+-ATPase.

Vegetative reproduction : Asexual reproduction of plants through detachment of a part that develops into a complete plant; clonal growth.

Vegetative storage protein : Proteins accumulating in vegetative plant parts (leaves and hypocotyls) at a high supply of nitrogen (e.g., in Glycine max).

Veins : The fine branches and intricate network of leaf vascular bundles.

Vernalin : Hormone responsible for vernalization effect.

Vernalization :

- A cold temperature treatment given to seeds or plants (belonging to biennial crops) that is needed for germination or for switching from vegetative stage to flowering stage (biennial crops).
- Acceleration of flowering ability by chilling treatment (Chourad,1960).
- In some species, the cold temperature requirement for flowering. The term is derived from the word for "spring".
- Induction of flowering by exposure to low temperatures (from the Latin word ver ¼ spring).

Vesicular arbuscular mycorrhizal fungi :

- The huphae grow in a less dense arrangement both within the root itself and extending outward from the root into the surrounding soil.
- A nondense mycelium with in the root itself and extending into the soil; some hyphae penetrate individual cells of the root cortex as well as extending through the regions between cells.

Vessel : Water-conducting element of the xylem

Vessel elements : Nonliving water conducting cells with perforated end walls found only in angiosperms and a small group of gymnosperms.

Vessel elements in Xylem : Stack of two or more Vessel elements which are dead cells and connected to one another through perforation plates.

Viability :

- Quality as state of being alive, ability to live, grown, and develop, as the viability of certain grains under dry conditions.
- State of aliveness of seed.

Viability equation : An equation to predict the viability of stored seeds based on behaviour of seeds to storage environment.

Viability test : A test used to measure whether the seed is alive or not, e.g. Terazolium test.

Viable :

- Capable of growing or developing, as viable seeds.
- Having the potential to germinate.

Vigour :

- Condition of active good heath and natural robustness, when seeds are planted vigour permits germination to proceed rapidly and to completion under a wide range of conditions.
- Seed vigour is the sum total of those properties of the seeds which determine the level of activity and performance of a non-dormant seed or seedlot during germination and seedling emergence (Perry, 1978).

Vigour tests : A group of physical and biochemical tests to assess the vigour of seedlot.

Viscoelastic : Creep mixture of viscous and elastic processes during cell-wall expansion; also unsavory character met in dark alleys.

Viscoelestic or rheological properties :

- Physical properties of cell wall that are intermediate between those of a solid and those of a liquid.
- Properties that are intermediate between those of a solid and those of a liquid and combine viscous and elastic behaviour.

Vital theories : Upward movement of water is due to pumping activity of cells of wood parenchyma and medullary rays due to periodic change in their osmotic pressure Godlewski (1884). Ascent of sap is due to pulsatory activity of cortical cells of stem just outside endodermis (J.C. Bose, 1923).

Vitamins : Oranganic substanes produced by plants act in association with enzymes as prosthetic group and enhance their activity. Site of synthesis and action are same.

Vitrification : Procedure for long term storage at ultra low temperature in which samples (tissue or embryos) are placed for pretreatment in extremely concentrated cryoprotective solutions, then freezed ultra rapidly (in liquid nitrogen) in other to avoid formation of intracellular ice crystals.

Vitro : Literally in glass, in a test tube, bottle etc.

Viviparous seeds : Seeds that germinate prior to abscission from the maternal plant (e.g., mangrove species).

Vivipary :

- The germination of seeds in the mother plant itself without being dried. **e.g.** Chow- chow (*Sechium edule*); Mangrove tree (*Rhizophora* Sp.).
- The precocious germination of seeds in the fruit while still attached to the plant.

Vivo : In situ, in the intact plant.

VLFR(very low fluence response) : Phytochrome responses whose magnitude is proportional to very low fluence (1 to 100 nmol m^{-2}).

Vmax : Substrate-saturated rate of a chemical conversion catalyzed by an enzyme (expressed per unit protein, rather than per mole catalytic sites as in kcat).

❑❑❑

Wall creep : In isolated primary cell walls, the time dependent irreversible extension of cell wall polymers due to their slippage relative to one another.

Wall extensibility (m) : During primary cell wall expansion, the coefficient that relates growth rate to the turgor pressure that is in excess of the yield threshold.

Wall loosening : Refers to the process during which covalent or noncovalent bonds between cellulose microfibrils and other macromolecules are broken, so that the cell under turgor can expand.

Warburg effect : Inhibition of photosynthetic rate in presence of O_2.

Water channel : Pore for water transport in membranes consisting of a specialized protein (aquaporin); water moves single file.

Water core development in Turnip : Boron deficiency caused rotting of skin in turnip due to increase in water content.

Water curtain germinator : A germinator that is heated, cooled, aerated and humidified by means of a curtain of water falling down the back of germinator.

Water deficit :

- Any water content of a plant cell or tissue below its water content when fully hydrated.
- Any water content of a tissue of cell or tissue that is below the highest water content exhibited at the most hydrated state.

Water lily type stomata : Only on upper surface. Aquatic plants with floating leaves. Example-Water lily etc.

Water potential :

- Water potential is a measure of the free energy associated with water per unit volume (J m^{-3}). These units are equivalent to pressure units such as pascals. Ψ_w is a function of the solute potential, the pressure potential and the potential due to gravity : $\Psi_w = \Psi_s + \Psi_p + \Psi_g$ The term Ψ_g is often ignored because it is negligible for heights under five meters.
- Chemical potential of water in a system or part of system minus chemical potential of pure water at the same atmospheric pressure, temperature and height represented with Ψ Psi (Electrical potential). The components are solute or osmotic potential, matric potential and pressure potential. It is represented with (-) sign.
- Chemical potential of water divided by the molar volume of water, relative to that of pure water at standard temperature and pressure.

Water potential difference ($\ddot{A}Y_w$) : The water potential difference across the plasma membrane (expressed in megapascals).

Water status : Loose term referring to aspects of the plant's relative water content, turgor, water potential, etc.

Water stress : Stress due to shortage of water.

Water table : At some depth in soil all the pore spaces are filled with water. If a hole is bored in the soil water will appear at this point.

Water-use efficiency :

- Ratio between the gain of (above-ground) biomass in growth or CO_2 in photosynthesis and transpirational water loss.
- Water use efficiency is reciprocal of transpiration rate.

$$\frac{1 \quad CO_2}{500 \text{ Water transpired}} = 0.002$$

Watery seedling : A form of abnormal seedling in monocots where the coleoptile sheath is filled with water. **e.g.** Maize.

Wavelength (ë) : A unit of measurement for characterization of light energy. The distance between successive wave crests. In the visible spectrum, corresponds to a colour.

Waxes :

- Esters of fatty acid and a monohydric alcohol found in seed coats, solids at room temperature.
- Complex mixtures of long chain acyl lipids that are extremely hydrophobic. The most common compounds of waxes are straight chain alkanes and alcohals of 25 to 35 carbon carbon atoms. The waxes of cuticle are synthesized by epidernal cells.
- Esters of high molecular weight monohydric alcohals and high molecular weight fatty acids.
- Complex mixtures of extremely hydrophobic lipids synthesized and secreted by epidermal cell and that make up the protective cuticle that reduces water loss from exposed plant tissue. Waxes are mostly alkanes and alcohols of 25 to 35 carbon atoms.

Weak electrolytes : Substances which poorly dissociate when dissolved. Examples - acetic acid etc.

Weathering damage : Damage to seeds while on mother plant caused by exposure to high humidity; temperature; infection by fungi, insects, UV radiation.

Weed seed : Seed, bulblet, tuber of plants recognized as weed as per seed law.

Wetlands : Land that is saturated with water on a regular basis.

Whip tail disease : Due to molybdenum deficiency leaves show interveinal mottling, margin becomes grey and flacid and finally brown. The leaves tissue collapses leaving only midrib and small pieces of leaf blade which appears like a whip or tail.

White tag : Refers to foundation seed tag.

Whorled phyllotaxy : Several leaves or floral organs of a given type are attached to the stem at the same node.

Whorls : Set of the organs initiated in concentric rings.

Wilting :

- Drooping and sagging of plant tissues, specially leaves due to change' in elestic properties of cell walls when turgor pressure declines below a certain critical value.
- Plant loss of rigidity, leading to a flaccid state, due to turgor pressure falling to zero.

Wilting coefficient or permanent wilting point : Amount of moisture left in soil after a plant has permanently wilted. Lowest for sandy, high for loam and very high for clay soils.

Wilting point : Water potential at which turgor pressure is zero.

Winter annual species : Whose seeds germinate before or in winter and completes its life cycle before the start of the next summer; typically found in Mediterranean-type climates.

Working sample : The reduced sample got by maxing and dividing of the submitted sample in the seed testing lab; used for performing tests on seeds quality.

Wound callose : Callose deposited in the sieve pores of damaged sieve elements that seals them off from surrounding intact tissue. As sieve elements recover, the callose disappears from pores.

❑❑❑

X

Xanthophyll cycle :

- Chemical transformations of a number of carotenoid molecules in the chloroplast that avoid serious damage by excess radiation.
- The interconvertion of violaxanthin and zeaxanthin via the intermediate antheraxanthin. Zeaxanthin accumulates under conditions of excess energy.

Xanthophylls : Carotenoids involved in nonphotochemical quenching. The xanthophyll zeaxanthin is associated with the quenched state of photosystem II, and violaxanthin is associated with the unquenched state.

Xenobiotic : Potentially toxic chemical that is found in an organism where it is normally not occurring; can be restricted to synthetic compounds, but is also used in a wider sense.

Xerophyte :

- Plant structurally adapted for growth with a limited water supply.
- Plant that typically grows in dry Environments.

Xylem :

- Consists of vessel elements, Xylem (wood) parenchyma, fibres and tracheids.
- The vascular tissue that transports water and ions from the root to the other parts of the plant.

Xylem loading : The process whereby ions exit the symplast and enter the conducting cells of the xylem.

Xylem tracheids : Elongated, hollow dead cells with highly lignified walls.

Xyloglucan : A hemicellulose with a backbone of 1 4 linked β-D - glucose residues and short side chains that contain xylose, galactose and sometimes fucose. It is the most abundant hemicellulose in the primary walls of most plants (in grasses it is present, but less abundant).

□□□

Yield coefficient : A proportionality constant in the Lockhart equation that refers to the plasticity of cell walls.

Yield threshold (Y) :

- The minimum value for turgor pressure at which measurable extension of the cell wall begins.
- Minimum turgor pressure for cell Expansion.

Yielding : Long term irreversible stretching that is characteristic of growing (expanding) cell walls. Nearly lacking in nonexpanding walls.

❏❏❏

Z

Z ("zigzag") scheme : Arrangement of the reaction center and antenna complexes of photosystem II and Photosystem I, and the electron transport chain that links them by their midpoint redox potential. The resulting alignment is shaped like a letter "Z".

Zeatin (6-(4 hydroxy-3- methyl-but-trans-2-enyl)-aminopurine) :

- A naturally occurring cytokinin found in maize kernels.
- A naturally occurring cytokinin that stimulates mature plant cells to divide when added to a culture medium along with an auxin. Chemically, trans-6 (4-hydroxy-3-methylbut-2enylamino) purine.
- A phytohormone belonging to the cytokinins, the name stems from Zea mays (corn), from which it was first isolated.
- Substance stimulates plant cells to divide when added to a culture medium alongwith an auxin identified as trans 6 (4 - hydroxy - 3 - methylbut - 2 - enylamino) purine which was named as zeatin by Letham (1973).

Zeatin ribosides : Zeatin with a ribose attached to the amino purine moiety. The main cytokinin in the xylem exudate.

Zeaxanthin : A carotenoid implicated as a blue light photoreceptor. A component of the xanthophyll cycle of chloroplasts, which protects from excess excitation energy.

Zeitgebers : Environmental signals such as light to dark or dark to light transitions that synchronize the endogenous oscillator to a 24 hour periodicity.

Zone of elongation : Part of root where growth takes place in length having slow absorption rate.

Zygote :

- Produced by fusion of one sperm nuclei with egg cell or egg nucleus , containing 2n chromosome set connected with storage of food material. A fertilised egg.
- The single celled product of the union of an egg and a sperm.

□□□

About the Authors

Dr. A.S. Gontia, Professor, Department of Plant Physiology, JNKVV, Jabalpur was born on 29th April 1957. He has completed his graduation from the University of Jabalpur during 1976 and obtained the M.Sc. (Ag.) degree in Agricultural Botany during 1979. He obtained his Ph.D. degree in Botany from the APS University Rewa in the year 1990. He joined his services during 1980 in the Department of Agricultural Botany which was later on redesignated as Deptt. of Plant Physiology as Assistant Professor. During 2001 he was promoted to the post of Professor. He has been actively involved in teaching, research and extension and guided 15 M.Sc. (Ag.) students as major advisor and 8 Ph.D. students of various disciplines including Plant Physiology as minor advisor. His major field of expertise is Crop Physiology and Medicinal and Aromatic Plants. He has 42 research papers, book chapters, bulletins to his credit. Presently besides teaching of UG, PG and Ph.D. students he is involved in research and extension activities pertaining to crop plants as well as medicinal and aromatic plants.

Dr. (Mrs.) Sathrupa Rao, Professor and Head, Department of Plant Physiology, JNKVV, Jabalpur completed her graduation from University of Jabalpur and post graduation M.Sc. (Ag.) in Crop Physiology from Jawaharlal Nehru Krishi Vishwa Vidyalaya during 1970 and 1974. She joined her services during 1978 as Senior Research Assistant and has rendered her services in various capacities. She has been actively involved in teaching, research and extension and has guided around 25 M.Sc. (Ag.) and one Ph.D. students in various aspect of Plant Physiology. Her major field of expertise is in Seed physiology, Storage and Testing. She has 30 research articles, book chapters, bulletins and manuals to her credit. Dr. Sathrupa is a recipient of ICAR award for "Outstanding Interdisciplinary Team Research" for her contribution in seed physiological aspects of Soybean. Presently she is involved in research and training activities related to medicinal and aromatic plants.